NMR and Biomolecular Structure

Edited by
I. Bertini, H. Molinari, N. Niccolai

VCH

Distribution:
VCH, P. O. Box 101161, D-6940 Weinheim (Federal Republic of Germany)
Switzerland: VCH, P. O. Box, CH-4020 Basel (Switzerland)
United Kingdom and Ireland: VCH (UK) Ltd., 8 Wellington Court, Cambridge CB1 1HZ (England)
USA and Canada: VCH, Suite 909, 220 East 23rd Street, New York, NY 10010–4606 (USA)

ISBN 3-527-28222-X (VCH, Weinheim) ISBN 1-56081-116-1 (VCH, New York)

NMR and Biomolecular Structure

Edited by
Ivano Bertini, Henriette Molinari,
Neri Niccolai

VCH Weinheim · New York · Basel · Cambridge

Prof. Ivano Bertini
Department of Chemistry
University of Firenze
Via G. Capponi 7
50121 Firenze, Italy

Dr. Henriette Molinari
Department of Organic Chemistry
University of Milano
20133 Milano, Italy

Prof. Neri Niccolai
Department of Chemistry
University of Siena
53100 Siena, Italy

Published jointly by
VCH Verlagsgesellschaft mbH, Weinheim (Federal Republic of Germany)
VCH Publishers, Inc., New York, NY (USA)

Editorial Director: Dr. Michael G. Weller
Production Manager: Elke Littmann

Cover illustration: Non-selective 300 MHz ROESY-TOCSY spectrum of buserilin in DMSO-d_6, photographed from an Evans and Sutherland Picture System.

Library of Congress Card No. applied for.

A CIP catalogue record for this book is available from the British Library.

Deutsche Bibliothek Cataloguing-in-Publication Data:
NMR and biomolecular structure / ed. by Ivano Bertini ... –
Weinheim ; New York ; Basel ; Cambridge : VCH, 1991
ISBN 3-527-28222-X (Weinheim ...) Gb.
ISBN 1-56081-116-1 (New York) Gb.
NE: Bertini, Ivano [Hrsg.]

Printed on acid-free paper.

Printing: betz-druck gmbh, D-6100 Darmstadt. Bookbinding: Verlagsbuchbinderei Kränkl, D-6148 Heppenheim.
Printed in the Federal Republic of Germany

Preface

Every year at the end of July there is a horse race in Siena which is called the Palio. In 1988 some of the contributors to this volume attended the Palio on the occasion of a meeting on "Molecular Structure, Dynamics and Recognition" and were able to share the combined excitement of this NMR meeting with the Palio. In the field of NMR, horses and spins have long been associated since the famous slides shown by Ray Freeman and we were pleased to continue the tradition. At the time we thought about a multiauthor NMR book which would be didactical, providing an introduction to the frontier NMR techniques while at the same time containing some meaningful applications to macromolecular systems.

The first chapter deals with the basic Fourier transform operations and introduces two dimensional NMR for analyzing the scalar correlation. In detail the second introduces the consequences of spin relaxation in one and two dimensional correlations experiments. This is presented within the frame of a treatment of fast relaxing (paramagnetic) systems. In these two chapters both COSY and NOESY experiments are explained in a pictorial way. The third chapter introduces 3D experiments after providing a critical evaluation of the limits of 2D experiments. The various pairwise combinations of 2D schemes that make up a 3D sequence are discussed with the aim of showing how they achieve better resolution of complex spectra and of comparing the sensitivity of the various experiments. A complete relaxation matrix analysis is discussed in chapters four and five. This is necessary when using NOE measurements for quantitative distance analysis. Chapter four presents an application of this method to the structural characterization of DNA fragments.

In chapter five particular attention is given to the approach using molecular dynamics in combination with structural constraints obtained from NOE data. An application of this method to the structure determination of a DNA-protein complex is given. Chapter six and seven present elegant pieces of work in the study of structures and dynamic processes. The former deals with the application of ^{1}H and ^{31}P 2D spectroscopy to the elucidation of oligonucleotides in conjunction with molecular dynamics. The latter deals with the dynamics of the interactions of small molecules with proteins using 1D and 2D exchange spectroscopy with a critical discussion of the problem of multiple conformations.

We hope we have provided the NMR community with an interesting text: concise yet providing a good balance between theory and applications. The level is intended for graduate students in chemistry and biophysics. We were fortunate to have some of the best scientists in the area among the contributors ... and we are grateful to them for contributing to this volume.

January 1990

Ivano Bertini
Henriette Molinari
Neri Niccolai

Contents

List of Contributors

Lucia BANCI
Department of Chemistry
University of Firenze
Via G. Capponi 7
50121 Firenze, Italy.

Ivano BERTINI
Department of Chemistry
University of Firenze
Via G. Capponi 7
50121 Firenze, Italy.

Anna Maria BIANUCCI
Departments of Pharmaceutical Chemistry and Radiology
University of California
San Francisco, CA 94143 USA.

Rolf BOELENS
Department of Chemistry
University of Utrecht
Padualaan 8
3584 CH Utrecht, The Netherlands.

Brandan BORGIAS
Departments of Pharmaceutical Chemistry and Radiology
University of California
San Francisco, CA 94143 USA.

Gennaro ESPOSITO
Department of Biology
University of Udine
33100 Udine, Italy.

James FEENEY
National Institute for Medical Research
Mill Hill, London NW7 1AA, UK.

Josepha A.M. FU
Department of Chemistry
Purdue University
W. Lafayette, IN 47907 USA.

David G. GORENSTEIN
Department of Chemistry
Purdue University
W. Lafayette, IN 47907 USA.

Christian GRIESINGER
Department of Organic Chemistry
University of Frankfurt
Niederurseler Hang
D-6000 Frankfurt am Main, West Germany.

Thomas L. JAMES
Departments of Pharmaceutical Chemistry and Radiology
University of California
San Francisco, CA 94143 USA.

Claude R. JONES
Department of Chemistry
Purdue University
W. Lafayette, IN 47907 USA.

Robert KAPTEIN
Department of Chemistry
University of Utrecht
Padualaan 8
3584 CH Utrecht, The Netherlands.

Cristine KARSLAKE
Department of Chemistry
Purdue University
W. Lafayette, IN 47907 USA.

Thea M.G. KONING
Department of Chemistry
University of Utrecht
Padualaan 8
3584 CH Utrecht, The Netherlands.

Claudio LUCHINAT
Institute of Agricultural Chemistry
University of Bologna
Viale Berti Pichat 10
40127 Bologna, Italy.

Paolo MASCAGNI
Department of Pharmaceutical Chemistry
University of London
London WC1N-1AX, UK.

James T. METZ
Department of Chemistry
Purdue University
W. Lafayette, IN 47907 USA.

Henriette MOLINARI
Department of Organic Chemistry
University of Milano
20133 Milano, Italy.

Neri NICCOLAI
Department of Chemistry
University of Siena
53100 Siena, Italy.

Edward NIKONOWITZ
Department of Chemistry
Purdue University
W. Lafayette, IN 47907 USA.

Mario PICCIOLI
Department of Chemistry
University of Firenze
Via G. Capponi 7
50121 Firenze, Italy.

Robert POWERS
Department of Chemistry
Purdue University
W. Lafayette, IN 47907 USA.

Vikram A. ROONGTA
Department of Chemistry
Purdue University
W. Lafayette, IN 47907 USA.

Robert SANTINI
Department of Chemistry
Purdue University
W. Lafayette, IN 47907 USA.

Stephen A. SCHROEDER
Department of Chemistry
Purdue University
W. Lafayette, IN 47907 USA.

Ning ZHOU
Departments of Pharmaceutical Chemistry and Radiology
University of California
San Francisco, CA 94143 USA.

List of Abbreviations and Chemical Symbols

^{3}J	three bonds coupling constant
A	absorption Lorentzian lineshape
AMBER	assisted model building with energy refinement analysis of NMR data
B_0	static magnetic field
B_{eff}	effective magnetic field
COLOC	correlation spectroscopy via long range couplings
COMATOSE	complex matrix analysis torsion optimized structure
CORMA	complete relaxation matrix analysis
COSY	2D correlated spectroscopy
CW	continuous wave
D	dispersion Lorentzian lineshape
DDD	distance bounds driven dynamics
DG	distance geometry
DIRECT	direct calculation of distances
DISGEO	distance geometry
DISMAN	computer program for the structural analysis of NMR data
DMSO	dimethylsulfoxide
DNA	deoxyribonucleic acid
DOC	double constant time 2D experiment
DQF COSY	double quantum filter COSY
DTPA	diethylenetriaminepentaacetate
E	unity operator
EM	energy minimization
EXSY	exchange spectroscopy
F_1, F_2	frequency domains
FID	free induction decay

h	Planck's constant
HOHAHA	homonuclear Hartmann-Hahn
HRP-CN	horseradish peroxidase-CN
I	spin quantum number
I^+	raising operator
I^-	lowering operator
Im	imaginary part of the spectrum
IRMA	iterative relaxation matrix approach
ISPA	isolated spin pair approximation
I_x, I_y, I_z	angular momentum operators
$J(\omega)$	spectral density at frequency ω
J_{IS}	coupling constant between I and S nuclei
LW	line width
M_0	equilibrium macroscopic magnetization
MD	molecular dynamics
metMb-CN	metmyoglobin-CN
MM	molecular mechanics
MTX	methotrexate
MW	molecular weight
M_z, M_x, M_y	components of macroscopic magnetization
NADPH	nicotinamide adenine dinucleotide phosphate
NiSal-MeDPT	[NN'-4-Methyl-4-Azaeptane-1,7-diylbis (Salicylideneiminato)]Ni(II)
NOESY	2D nuclear Overhauser and exchange spectroscopy
PAC	pure absorption phase constant time
Pu	purine base
Py	pyrimidine base
Re	real part of the spectrum
REM	restrained energy minimisation
RMD	restrained molecular dynamics
RMS	root mean square
RNA	ribonucleic acid
ROESY	rotating frame NOE spectroscopy
$S(t_1, t_2)$	function of the two time domains
S/N	signal to noise ratio

SOD	superoxide dismutase
SW	sweep width
t_1	evolution time
t_2	acquisition time
T_1	longitudinal relaxation time
T_2	transverse relaxation time
t_g	global helical twist
TMP	trimethoprim
TOCSY	total correlation spectroscopy
w_0	zero quantum transition probability
w_1	single quantum transition probability
w_2	double quantum transition probability
XRD	X-Ray diffraction
α, β	spin eigenfunction of a nuclear spin I
γ	gyromagnetic ratio
Δt	time between sampling points
η_I	fractional change in the integrated NMR signal intensity
μ_0	permeability of a vacuum
ρ	intrinsic relaxation rate
σ	cross-relaxation rate
τ_c	isotropic correlation time
τ_s	electronic correlation time
τ_m	mixing time
τ_r	rotational correlation time
Ω	chemical shift expressed in radians
ω	Larmor frequency expressed in radians

1. Structural NMR Studies: from One to Multidimensional Frequency Spectra

Gennaro Esposito, Paolo Mascagni, Henriette Molinari and Neri Niccolai

1.1 Introduction

Since the first successful detection of an NMR signal, late in 1945 [1,2], nuclear magnetic resonance has developed in such a spectacular fashion that revolutionary new developments in instrumentation, data processing and pulse sequences are continuously being added to an already impressive wealth of NMR tools.

Much of the thrust for these developments has been derived from the application of NMR to biological problems. Nowadays, the combination of high field spectrometers and the most advanced multidimensional techniques is used to obtain structural information for biopolymers, that, only recently, in the late 60's were unthinkable even for concentrated solution of small organic molecules.

The evolution of NMR has been marked by the development of high field superconducting magnets as well as the advent of Fourier transformation methods. The improvement in sensitivity which followed the original work of Anderson and Ernst in 1966 [3], led to an improved sensitivity particularly for those less abundant but important nuclei such as ^{13}C and ^{15}N.

Another advantage achieved by pulsed FT-NMR was the increased ability to manipulate nuclear spins and thus generate new information.

Double resonance and double difference spectroscopies, longitudinal and transversal relaxation measurements are just a few of the many sophisticated experiments which were developed in the early 70's and throughout the decade.

However, the major breakthrough in recent years has followed a third development of the original Fourier transform concept: two-dimensional NMR spectroscopy. Jeener in 1971 [4] and Ernst and co-workers in 1976 [5] proposed and implemented the idea of two-dimensional Fourier transformation of NMR data. And the history of NMR repeated itself: sensitivity, resolution and ability to manipulate a nuclear spin all received a new thrust, only this time in a second (and third!) dimension.

The richness and diversity of pulse sequence libraries thus derived have been put to good use by the NMR spectroscopist. The combination of a suitably designed experimental strategy and the availability of high field spectrometers (typically in the 400-600 MHz range) has resulted in the elucidation of the chemical and 3D structure of many biopolymers with molecular weights up to about 10 KDaltons.

Here we try to give a contribution to improve the knowledge of young spectroscopists by outlining some theoretical aspects of FT-NMR and discussing experimental strategies and limitations.

1.1.1 The Fourier Transform (FT) Revolution

In conventional continuous-wave (CW) NMR spectroscopy most of the time required to measure the absorption spectrum is consumed by noise recording. This limited drastically the detection of weak resonances of dilute spin systems particularly in those cases where the observed nucleus had a low natural abundance and/or a small gyromagnetic ratio.

One of the modifications proposed to improve the spectral S/N ratio, i.e. multichannel spectrometers, required so severe instrumental changes that it never found application on industrial scale. However, in the late 60's, Ernst [3] showed that a short delta-function-like pulse could be regarded as a multi-frequency source for the simultaneous excitation of all resonance frequencies.

The immediate practical consequence of this theory was a dramatic improvement of the S/N ratio, the "Fellgett advantage", which, for a repeated pulse experiment, could be quantified as $N^{1/2} = (SW/LW)^{1/2}$, where N is the number of transients, SW is the observed sweep width and LW is the line width of the NMR signals. The introduction of pulsed NMR as a practical spectroscopy was also favoured by the evolution of computer systems and computational methods, for instance towards the end of the 60's, low-cost minicomputers could efficiently analyze the response of a radio frequency pulse by using the Cooley-Tukey algorithm for the fast Fourier transformation [6].

Since the fundamental aspects of multidimensional NMR are shared by one-dimensional Fourier Transform nuclear magnetic resonance (1D FT-NMR), we will recall here some of the basic concepts of the FT-NMR spectroscopy.

The Fourier transformation establishes the correspondence between the functions s(t) in the time domain and S(ω) or S(f) in the frequency domain:

$$S(\omega) = \int_{-\infty}^{\infty} s(t)\, e^{-i\omega t}\, dt$$

$$S(f) = \int_{-\infty}^{\infty} s(t)\, e^{-i2\pi ft}\, dt$$

$$s(t) = \frac{1}{2\pi} \int_{-\infty}^{\infty} S(\omega)\, e^{i\omega t}\, d\omega$$

$$= \int_{-\infty}^{\infty} S(f)\, e^{i2\pi ft}\, df$$

where $\omega = 2\pi f$. In Appendix some of the fundamental relationships between Fourier transformed pairs of functions, are indicated.

1.1.2 Mono-dimensional FT-NMR

When a nucleus with spin I = 1/2 is placed in a magnetic field of strength B_0, two energy levels are generated: a lower state, which is generally labelled α, and contains the nuclear magnetic moment parallel to the applied magnetic field B_0; the upper state (β) in which the magnetic moments are anti-parallel to B_0. The two energy states are unequally populated, the ratio of the populations being given by the Boltzmann distribution law. The population difference yields a macroscopic longitudinal magnetization whose behaviour can be described in terms of phenomenological Bloch equations [7]. The effects of high frequency oscillating magnetic fields of short duration (pulses) on the bulk magnetization can be conveniently described using vector diagrams (Fig. 1), in a reference rotating frame.

$$M_z \xrightarrow{\beta_y} M_z \cos\beta + M_x \sin\beta$$

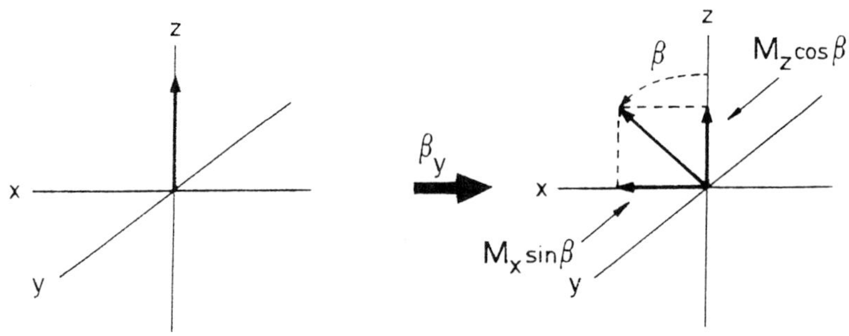

Fig. 1 Effect of pulses on the bulk magnetization.

Immediately after a torque is produced by the application of a pulse, the transverse magnetization, M_x, will precess around the z axis at an angular frequency $\omega = \gamma B_{eff}$. B_{eff} is the effective magnetic field experienced by each nucleus and depends on the nuclear shield; γ is the gyromagnetic ratio constant typical of the nucleus under investigation. The precessing magnetization (in the xy plane) will induce, in a tuned coil, an oscillating current (at the frequency ω) which, after amplification and mixing with a reference frequency (ω_o), can be detected. Owed to relaxation phenomena, the transverse magnetization then returns to equilibrium and the induced current decays. The detector is hence recording a signal $s(t_2)$ oscillating at a frequency $\omega - \omega_o = \Omega$ (chemical shift) which is decreasing in intensity during the time t_2. The signal is called free induction decay (FID).

The effect that the chemical shift has on the transverse magnetization is described by Fig. 2

$$M_x \xrightarrow{\Omega t} M_x \cos\Omega t + M_y \sin\Omega t$$

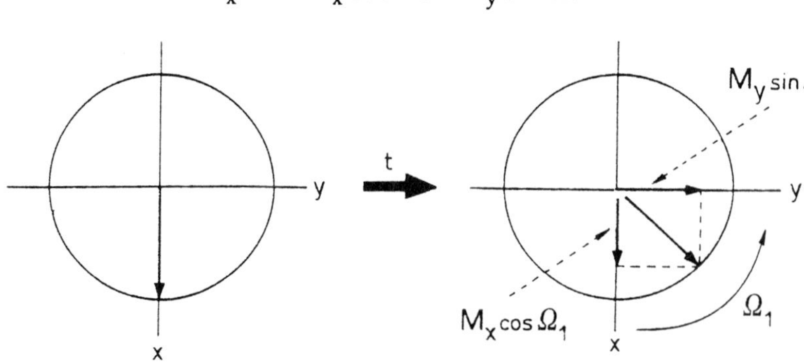

Fig. 2 Effect of the chemical shift on transverse magnetization.

Fourier transformation converts the signal $s(t_2)$ into a frequency spectrum where the signal is found at a frequency Ω. As shown in Fig. 2 identical responses arise on the detector from Ω and $-\Omega$. This problem is now routinely circumvented by the use of quadrature detection technique [8]. A computer manipulation (phase correction) of the spectrum is needed to obtain the desired lineshape for the resonances with $\Omega \neq 0$, as indicated by the shift theorem of the Fourier Transformation. However in the case of multidimensional NMR and unless an absolute value presentation of the signal is desired, only suitably designed pulse sequences can be used to obtain pure lineshape spectra.

1.2 From 1D to 2D NMR: Scalar Correlation Spectroscopy

The two main advantages brought about by the advent of Fourier transformation methods in NMR are the dramatic improvement in sensitivity and the possibility of studying time-dependent phenomena such as relaxation and chemical exchange phenomena.

However, although many of the possible coherences occur under the influence of pulses, only a few are associated with observable magnetizations, namely those obeying the selection rule $\Delta m_1 = +/-1$. On the other hand in 2D NMR spectroscopy the coherences precessing during the evolution period need not to be observed. This allows us to circumvent the limitations imposed by the selection rule.

Multiple pulse sequences are then created which allow the NMR spectroscopist considerable latitude in controlling the form of spectra. Following the effects of various different pulse sequences on a spin system can yield a great deal of structural and motional information not obtainable from simple mono dimensional spectra.

A detailed and rigorous description of a multi-spin system perturbated by the most complex 2D or 3D pulse sequences, is beyond the scope of this review.

In what follows bidimensional correlation spectroscopy will be briefly discussed to provide the basis for a more detailed analysis of the Fourier transformation in the second dimension and data presentation [9].

The concept of two-dimensional FT-NMR was introduced by Jeener in 1971 [4]. The original 2D experiment, for the simplest case of non-coupled spins, will be briefly described as an introduction to a more detailed discussion on the 2D Fourier transformation and data presentation. Jeener's original sequence can be divided into four distinct stages:

Preparation	Evolution	Mixing	Detection
	t_1		t_2

Initially the spin is allowed to relax to equilibrium during the preparation period at the end of which a $90°_{-y}$ pulse converts the longitudinal magnetization into transverse magnetization, along x axis (Fig. 3).

During the evolution period t_1 the transverse magnetization processes (Fig. 3b),

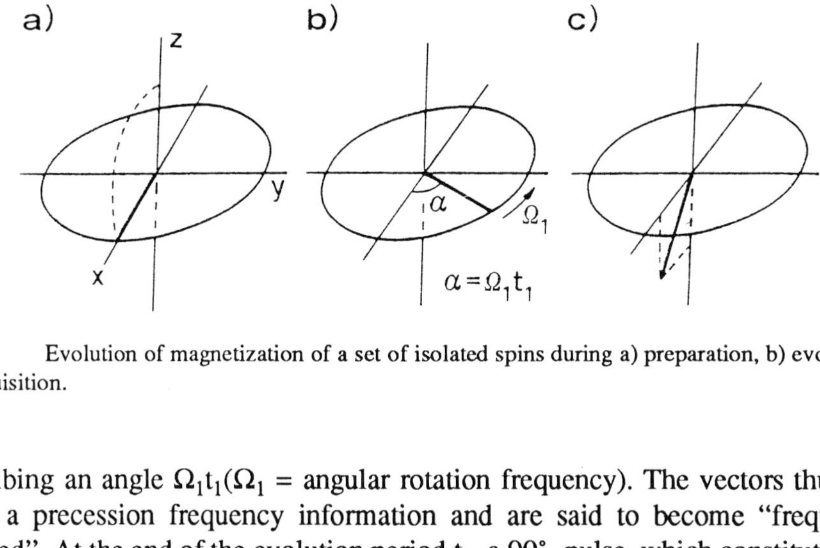

Fig. 3 Evolution of magnetization of a set of isolated spins during a) preparation, b) evolution, c) acquisition.

describing an angle $\Omega_1 t_1$(Ω_1 = angular rotation frequency). The vectors thus acquire a precession frequency information and are said to become "frequency labelled". At the end of the evolution period t_1, a $90°_x$ pulse, which constitutes the mixing period, is applied and a component of the magnetization vector is rotated back to the xz plane, while the other transverse component proportional to $\cos(\Omega_1 t_1)$ remains in the xy plane, parallel to the x axis.

A signal $s(t_1,t_2)$ is then measured during the detection period (Fig. 3c) as a function of the time t_2 and containing the time t_1 as parameter. If quadrature detection is employed during t_2 then:

$$s(t_1,t_2) = M_0\cos(\Omega_1 t_1)\ e^{i\Omega_2 t_2}\ e^{-t_1/T_2^{(1)}} e^{-t_2/T_2^{(2)}} \qquad (1)$$

where Ω_1 and Ω_2 are the precession frequencies during t_1 and t_2, respectively, and exponential relaxation is assumed to have time constants $T_2^{(1)}$ and $T_2^{(2)}$ during t_1 and t_2. In the simplest case of non-coupled spins, $\Omega_1 = \Omega_2$ and $T_2^{(1)} = T_2^{(2)}$.

The required Fourier transformation is that of a damped oscillator and, in general, the result of such a transformation is:

$$\mathrm{FT}\left\{e^{i\Omega_o t}\ e^{-t/T_2}\right\} = A + iD \tag{2}$$

where A is an absorption Lorentzian $A(\omega)$ and D is a dispersion Lorentzian $D(\omega)$, given by:

$$A(\omega) = \frac{T_2}{\left[1 + (\omega - \Omega_o)^2 T_2^2\right]} \tag{2a}$$

$$D(\omega) = \frac{(\omega - \Omega_o)T_2^2}{\left[1 + (\omega - \Omega_o)^2 T_2^2\right]} \tag{2b}$$

If t_1 is incremented in a stepwise fashion, Fourier transformation of the signal $s(t_1,t_2)$ with respect to t_2 will give an array of spectra proportional in amplitude to $\cos(\Omega_1 t_1)$:

$$s(t_1,\omega_2) = M_0\cos(\Omega_1 t_1)\ [A_2(\omega_2) + iD_2(\omega_2)]\ e^{(-t_1/T_2^{(1)})} \tag{3}$$

The absorptive parts of a set of these spectra, is shown in Fig. 4a. Cross sections parallel to the t_1 axis show a cosine modulation with angular frequency Ω_1, when $\omega_2 = \Omega_2$, and consist of ca. zero intensity for other ω_2 values. Usually the data matrix is transposed, in computer memory, to facilitate the second (Fig. 4b) Fourier transformation with respect to t_1. The FID's thus obtained are generally referred to as interferograms [10] and their Fourier transformation generates the final 2D spectrum (Fig. 4c): to facilitate their interpretation 2D spectra are usually and more conveniently presented as contour plots (Fig. 4d).

1.2.1 Sampling Frequency and Sensitivity

In normal quadrature detection the spectral width is $1/(2\Delta t)$ (or $\pm\ 1/(2\Delta t)$ in the case of simultaneous sampling with complex FT) where Δt is defined as the time between sampling points [11].

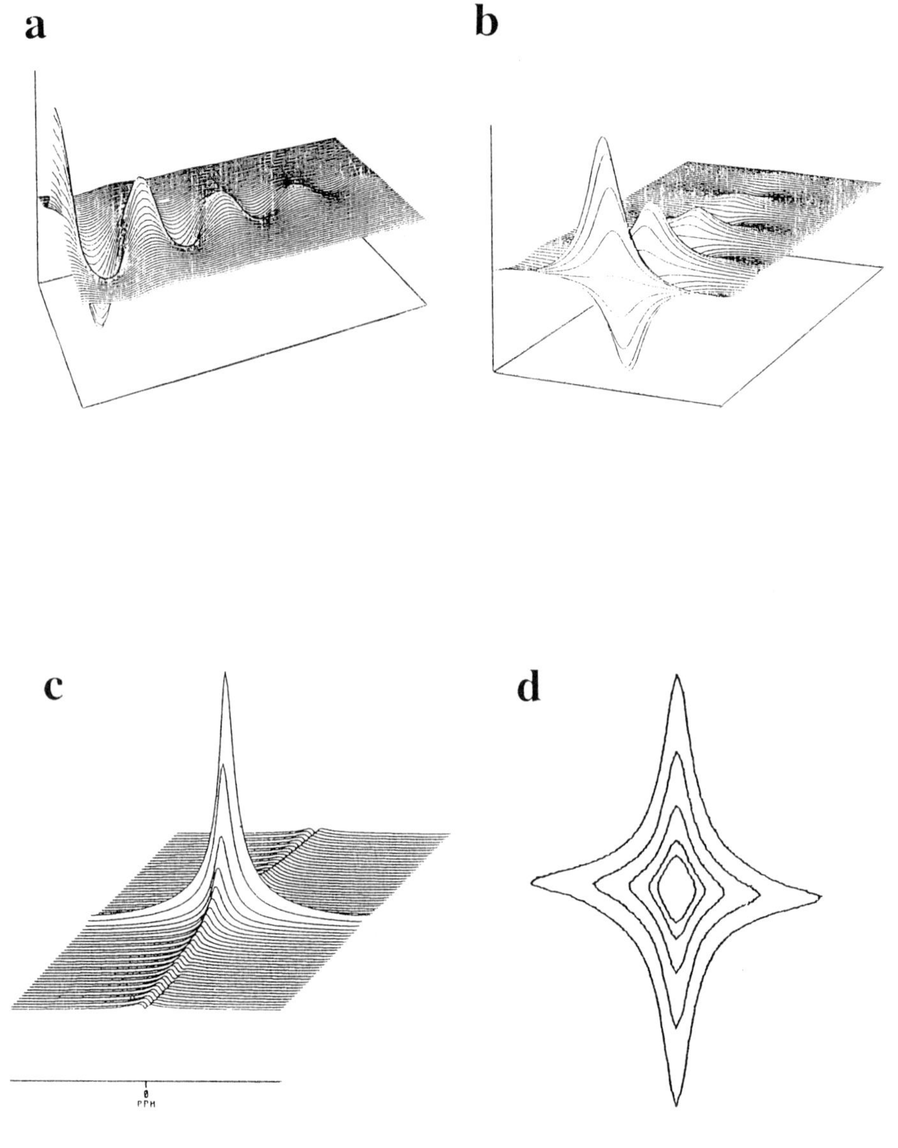

Fig. 4 a) absorptive part of a set of spectra described by eqn. (3) for increasing durations of a t_1 b) transposed data matrix of Fig. 4a c) final 2D spectrum: stacked plot d) final 2D spectrum: contour plot.

Likewise if consecutive t_1 values are separated by an amount Δt_1, the spectral width after the second Fourier transformation will equal $1/(2\Delta t_1)$. In general, Fourier transformation converts the signal energy of all data points into one (or more) narrow resonance lines. Similarly, the Fourier transformation with respect to t_1 combines the signal energy of a given resonance line from all the spectra collected at different t_1's and concentrates it into a single line of the 2D spectrum. Therefore the sensitivity of a 2D experiment needs not necessarily be lower than that of a 1D experiment. When sensitivity is a crucial problem the following suggestions [12,13,14,15] should be taken into account:

i) the acquisition time in the t_2 dimension should be at least $1.5 \times T_2$.

ii) the acquisition time in the t_1 dimension should not be longer than that absolutely necessary to obtain enough resolution in ω_1. The signal decays as a function of time t_1, hence spectra acquired for short t_1 values contribute the 2D resonance intensity more than those obtained for $t_1 \approx T_2^{(1)}$.

1.2.2 Lineshape and Frequency Discrimination

An amplitude cosine-modulated signal from a 2D experiment can be written as shown in eqn. (3). The time domain function of eqn. (3) contains no information as to the sign of Ω_1. This can be easily demonstrated [16] using the identities $\cos(x) = {}^1/_2[e^{ix} + e^{-ix}]$, and $\sin(x) = -{}^1/_{2i}\,[e^{ix} - e^{-ix}]$. Eqn. (3) may be rewritten as:

$$s_c(t_1,\omega_2) = \tfrac{1}{2}M_0(e^{i\Omega_1 t_1} + e^{-i\Omega_1 t_1})[A_2(\omega_2) + iD_2(\omega_2)]e^{(-t_1/T_2^{(1)})} \quad (3)$$

This after the second Fourier transformation with respect to t_1 becomes:

$$S_c\,(\omega_1,\omega_2) = \tfrac{1}{2}M_0\left[A_1^+(\omega_1) + iD_1^+(\omega_1) + A_1^-(\omega_1) + iD_1^-(\omega_1)\right]\left[A_2(\omega_2) + iD_2(\omega_2)\right] \quad (4)$$

where the notation $A_1{}^+D_1{}^+$ and $A_1{}^-D_1{}^-$ are the lineshape of peaks occurring at $\pm\,\Omega_1$ frequency. The real part of the spectrum $S_c(\omega_1,\omega_2)$ is:

$$\begin{aligned} \mathrm{Re}\left[S_c(\omega_1,\omega_2)\right] = \tfrac{1}{2}M_0\Big(&\left[A_1^+(\omega_1)\,A_2(\omega_2) - D_1^+(\omega_1)\,D_2(\omega_2)\right] + \\ &+ \left[A_1^-\,(\omega_1)\,A_2(\omega_2) - D_1^-\,(\omega_1)\,D_2(\omega_2)\right]\Big) \end{aligned} \quad (5)$$

Each term in parentheses on the right of eqn. 5 represents a phase-twist lineshape; it is not possible to assess from the second FT the sign of the modulation frequency Ω_1.

In general, to avoid confusion and to ensure that the modulation frequencies have all the same sign, the transmitter frequency is placed at one side of the spectral region. This method generates "pure-phase" spectra at the expense of heavy data handling and storage penalty. In addition, radio frequency offset effects may prove too severe in certain types of experiments.

One method to avoid these problems consists of introducing an artificial phase modulation. This is achieved by performing a second set of experiments 90_{-x}-t_1-90_x-acquire(t_2), where the required coherence, evolving during t_1, is phase-shifted by $\pi/2$ radians: the transverse magnetization is parallel to the x axis (at $t_2 = 0$) and its detected amplitude is modulated as $\sin(\Omega_1 t_1)$:

$$s_s(t_1,t_2) = M_0 \sin(\Omega_1 t_1)\, e^{i(\Omega_2 t_2 + \pi/2)}\, e^{-t_1/T_2^{(1)}}\, e^{-t_2/T_2^{(2)}} =$$

$$= iM_0 \sin(\Omega_1 t_1)\, e^{i\Omega_2 t_2}\, e^{-t_1/T_2^{(1)}}\, e^{-t_2/T_2^{(2)}} \qquad (6)$$

Combining eqns. (1) and (6) and omitting the relaxation terms gives

$$s^+(t_1,t_2) = M_0\, e^{(i\,\Omega_1\, t_1)}\, e^{(i\,\Omega_2\, t_2)} \qquad (7)$$

Eqn. (7) represents a signal whose phase at time $t_2 = 0$ is a linear function of the t_1 duration. Fourier transformation with respect to t_2 gives a resonance at frequency $\omega_2 = \Omega_2$, with phase $\Omega_1 t_1$:

$$s^+(t_1,\omega_2) = M_0\Big[\cos(\Omega_1 t_1)\, A_2(\omega_2) - \sin(\Omega_1\, t_1)\, D_2(\omega_2)\Big] +$$

$$+ iM_0\Big[\cos(\Omega_1 t_1)\, D_2(\omega_2) + \sin(\Omega_1\, t_1)\, A_2(\omega_2)\Big] \qquad (8)$$

A graphic representation of a set of spectra thus obtained is shown in Fig. 5. An interferogram taken at $\omega_2 = \Omega_2$ does not contain dispersive components since $D_2(\Omega_2) = 0$, and it is described by:

$$s^+(t_1,\Omega_2) = M_0\Big[\cos(\Omega_1 t_1) + i\sin(\Omega_1 t_1)\Big]\, T_2^{(2)} \qquad (9)$$

where the real part represents the interferogram taken through the real part of eqn. (7), and the imaginary part represents the interferogram taken through the imaginary parts of the $s^+(t_1, \Omega_2)$ spectra (Fig. 5b).

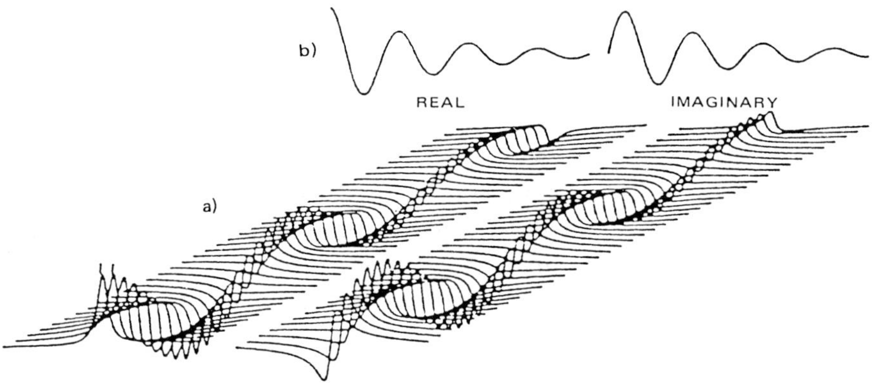

Fig. 5 Set of spectra represented by eqn. (8). *Reprinted from Ref. 9.*

The complex Fourier transformation of eqn. (9) leads to the determination of the sign of the modulation frequency Ω_1. Any interferogram at $\omega_2 \neq \Omega_2$, will have a non-zero contribution from $D_2(\omega_2)$ and its Fourier transformation will yield a signal, in the F1 dimension, 90 degree out of phase with respect to the interferogram at $\omega_2 = \Omega_2$. The complete 2D Fourier transformation of eqn. (7), obtained from the combination of the sine and cosine modulated signals, is

$$S^+(\omega_1,\omega_2) = M_0\left[A_1(\omega_1)A_2(\omega_2) - D_1(\omega_1)\,D_2(\omega_2)\right] +$$
$$+\, iM_0\left[A_1(\omega_1)\,D_2(\omega_2) + D_1(\omega_1)\,A_2(\omega_2)\right] \tag{10}$$

The real part of the spectrum, represented in Fig. 6, is

$$\mathrm{Re}\left[S^+(\omega_1,\omega_2)\right] = M_0\left[A_1(\omega_1)\,A_2(\omega_2) - D_1(\omega_1)\,D_2(\omega_2)\right] \tag{11}$$

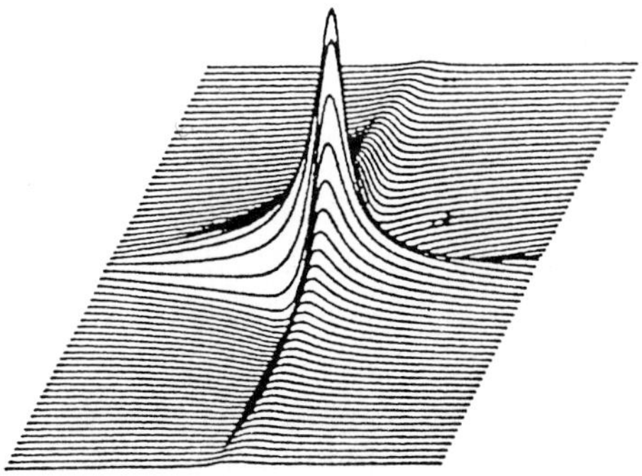

Fig. 6 Real part of eqn. (10). *Reprinted from Ref. 9.*

The sign of Ω_1 has thus been determined, although one unfavourable phase-twist lineshape has also been introduced. This signal cannot be phased in the pure absorption mode and absolute value mode calculation is usually performed:

$$\text{Absolute value} = \left[\text{Re}^2 + \text{Im}^2\right]^{1/2}$$

This resonance shows undesirable tails which decrease the spectral resolution. They can be suppressed by using appropriate digital filtering which reshapes the envelope amplitude of the time domain function [17]. Filters commonly used are the sine bell [18], the pseudo-echo window [19] and the convolution difference [20]. In Fig. 7 a comparison between two contour plots is shown: a) regular absolute value mode, b) the signal has been reshaped with a sine bell function in both dimensions.

Fig. 7 *Reprinted from Ref. 9.*

1.2.3 F1 Quadrature Detection and Absorption Mode

The described approach of converting the modulation of the signals during t_1 to one of phase rather than amplitude carries with it all the drawbacks described, and is therefore necessary to obtain ω_1 discrimination in some alternative way which avoids phase modulation.

Two methods have recently and independently been developed. The first of the two methods generates data suitable for a complex Fourier transformation with respect to t_1 and was originally described for a NOESY experiment [21]. The second, due to Marion and Wüthrich [22], requires a real Fourier transformation with respect to t_1 and was first applied to a COSY experiment.

To describe the first method in detail, let us consider again the pulse sequence of Fig. 3 and the signal of eqn. (1). A second experiment 90_{-x}-t_1-90_x-acquire(t_2) is performed. The signal thus generated is initially ($t_2 = 0$) along the x axis, but is modulated in amplitude as $\sin(\Omega_1 t_1)$. To obtain both frequency discriminations and pure two dimensional lineshapes the two signals $s_c(t_1,t_2)$ and $s_s(t_1,t_2)$ must be kept separate, rather than being combined during acquisition. By applying to the sine modulated signal the same processing procedure applied to the cosine modulated signal, another spectrum, consisting of two absorption-mode peaks can be generated. However, since the sine function is odd, one of the two peaks will be inverted.

If the two spectra $s_c(\omega_1,\omega_2)$ and $s_s(\omega_1,\omega_2)$ are then combined one of the peaks will cancel, giving frequency discrimination while retaining a pure absorption lineshape for the remaining peak. This can be easily demonstrated.
Fourier transformation of the two sets of spectra, with respect to t_2, gives:

$$s_c(t_1,\omega_2) = M_0 \cos(\Omega_1\, t_1)\left[A_2(\omega_2) + iD_2(\omega_2)\right] \tag{13a}$$

$$s'_s(t_1,\omega_2) = M_0 \sin(\Omega_1\, t_1)\left[A_2(\omega_2) + iD_2(\omega_2)\right] \tag{13b}$$

Now the imaginary part of $s_c(t_1,\omega_2)$ is replaced by the real part of $s'_s(t_1,\omega_2)$, yielding:

$$\begin{aligned} s^t(t_1,\omega_2) &= \mathrm{Re}\left[s_c(t_1,\omega_2)\right] + i\mathrm{Re}\left[s_s(t_1,\omega_2)\right] \\ &= M_0\left[\cos(\Omega_1\, t_1)\right]A_2(\omega_2) + iM_0\left[\sin(\Omega_1\, t_1)\right]A_2(\omega_2) \\ &= M_0\, e^{(i\Omega_1 t_1)}\, A_2(\omega_2) \end{aligned} \tag{14}$$

which, upon complex Fourier transformation with respect to t_1, gives for the real part:

$$S^t(\omega_1,\omega_2) = M_0\, A_1(\omega_1)\, A_2(\omega_2) \tag{15}$$

This represents a 2D absorption mode resonance.

The complete process is illustrated in Fig. 8.

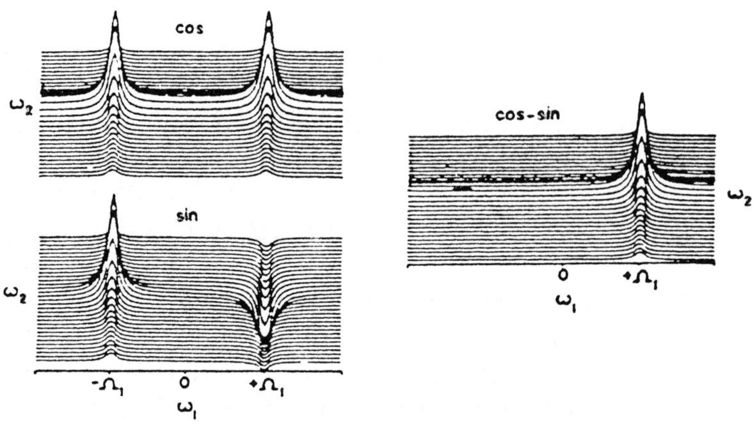

Fig. 8 Graphic representation of the process described by eqns. (13a-15). *Reprinted from Ref. 16.*

All that is required in this method is that for each t_1 value the sine and cosine modulated components are stored separately. Extra experiment time is not required since these two components are recorded with half the transients each and the full effect of time averaging is regained when the two signals are recombined.

The method used by Marion and Wüthrich to achieve the same result makes use of a real Fourier transformation. It is based on Redfield method [23] for obtaining quadrature detection in F_2.

In normal quadrature acquisition, a pair of data point, one from each quadrature channel, is sampled at intervals 1/SW. These two data points are considered as a complex time domain point and a spectrum is obtained after complex Fourier transformation.

In the Redfield modification, single data points are sampled at twice this rate and the phase of the receiver is advanced by $\pi/2$ radians after each data point is taken. For this reason the method is referred to as time-proportional phase incrementation, or TPPI. The spectrum is then obtained by performing a real Fourier transformation. The overall effect of this procedure is to add a frequency SW/2 to each point of the spectrum, so that peaks with offsets between –SW/2 and 0 will appear between 0 and SW/2, and peaks with offsets between 0 and SW/2 will appear between SW/2 and SW. In this way aliasing about zero frequency is prevented and quadrature detection is achieved. The same procedure can be applied to ω_1 dimension. The time t_1 is incremented in intervals of $1/(2SW_1)$ where SW_1 is the required sweep width in F_1 dimension. For each t_1 increment, the phase of the desired coherence evolving during t_1 is shifted by $\pi/2$ radians.

Consider, for example, the time domain signal $A_2(\omega_2)(\cos\Omega_1 t_1)$; the effect of the phase incrementation gives:

$$s(t_1,\omega_2) = \left[\cos(\Omega_1 t_1 + \pi t_1/2\delta)\right] A_2(\omega_2) \tag{16}$$

where δ is the sampling interval in the t_1 dimension and is equal to $1/(2SW_1)$. Eqn. 16 can be rewritten as:

$$s(t_1,\omega_2) = \left[\cos(\Omega_1 t_1 + \Omega_c t_1)\right] A_2(\omega_2) \tag{16a}$$

where $\Omega_c = 1/2(2\pi SW_1)$; in other words half of the spectral width has been added to the frequency of each line. As before, this avoids aliasing of signals in the range $+/-SW_{1/2}$ and an absorption mode lineshape is retained.

Although the relationship between this scheme and that of States is not obvious, a closer analysis [16] shows that the two methods are almost identical. Also from a practical point of view there is little difference: identical size data matrices are required and identical acquisition times are needed for the two methods, giving the same resolution and sensitivity in the final spectrum.

In conclusion the only difference between the two methods refers to the type of Fourier transformation applied with respect to t_1, which in turn depends on the type of spectrometer available.

1.3 Correlation in 2D Spectra

The results of the Jeneer 2D experiments on a system formed by non-coupled nuclear spins are easily accounted for using a simple vector picture, the Bloch equations, some basic multiplex spectroscopy concepts and elementary Fourier analysis mathematics. The desired modulation and phase characteristics of the signal can be selected by a proper choice of the data storage and processing procedure.

The next step forward in 2D NMR theory is the understanding of the origin of 2D correlation, i.e. the pattern observed in the presence of scalar or dipolar interactions, operating through chemical bonds or through space, respectively. Let's consider a two non-equivalent nuclei spin system (I = S = 1/2), in a static magnetic field.

For the time being, we focus our discussion on the scalar correlation 2D experiments, since the dipolar correlation experiments will be treated in detail in another section. Assuming that a resolved spin-spin coupling constant (J_{IS}) exists between the nuclei I and S, an experiment of 2D chemical shift correlation through scalar coupling can be performed, namely COSY (correlation spectroscopy), DQF COSY (double quantum filtered COSY), TOCSY (total correlation spectroscopy), etc. In general, the kind of response sorted out by these experiments is determined by their specific pulse sequence and phase cycle, and thus, their suitability depends on the characteristics of the system under investigation.

For our simple spin system, however, all the above mentioned experiments give the same information and provide insights in 2D correlation. Therefore, to illustrate the latter point, we prefer to use the simplest scheme, i.e. the COSY experiment, executed in the following manner:

preparation	evolution	mixing	acquisition
$90°_x$	---- t_1 ----	$90°_x$	t_2

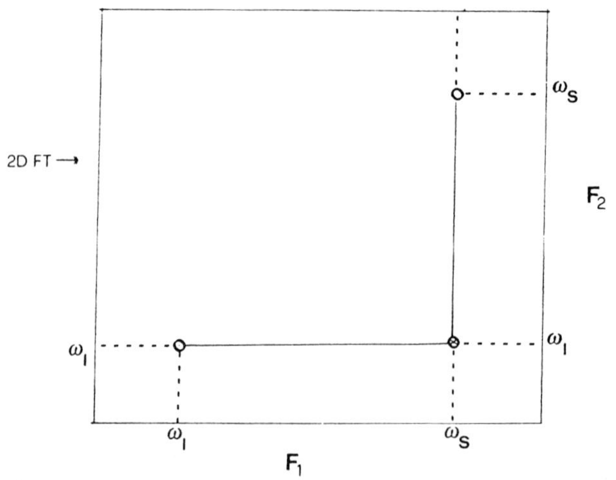

Fig. 9 Schematic contour plot presentation of a 2D NMR spectrum.

Two non-selective 90° radio frequency pulses are applied along the x axis, separated by a regularly incrementing delay, Δt_1, and followed by data acquisition during t_2. The step size of the t_1 increments determines the F_1 spectral width $SW_1 = 1/(2\Delta t_1)$.

We do not care, at this stage, about the kind of lineshape that the phase cycling scheme, the data storage and the processing protocol will produce.

A two dimensional Fourier transformation then gives a two dimension NMR spectrum, containing diagonal peaks, (those at the same frequency on both frequency axes) and cross-peaks (exhibiting different frequencies on the frequency axes). In the schematic contour plot presentation above depicted, diagonal peaks are indicated by filled circles, cross-peaks by empty circles. The presence of a symmetric pair of cross-peaks in a COSY contour plot indicates that a spin-spin coupling constant connects the two corresponding resonances. Now we want to know where the diagonal and cross-peaks come from.

1.3.1 Quantum Mechanical Description

A simple vector picture is of no help for this purpose, because it can only describe the behaviour of macroscopic observables, i.e. magnetization, whilst here we need to follow coherences, that behave according to the laws of quantum mechanics.

A proper treatment of the phenomena, taking place whenever a spin system in a static magnetic field is subjected to perturbative radio frequency pulses, can be accomplished using the spin density operator approach [24,25]. For a coupled two nuclei spin system (I = S = 1/2) the usual diagram

$\beta_I\beta_S$ $\qquad\qquad$ $m^{\alpha} = +1/2$

$\beta_I\alpha_S$ $\qquad$ $\alpha_I\beta_S$ $\qquad\qquad$ $m^{\beta} = -1/2$

$\alpha_I\alpha_S$ $\qquad\qquad$ (for positive gyromagnetic ratios)

describing all the possible transitions and populations can be accounted for by a matricial operator, named density matrix, ρ, defined as:

$$\rho = |\varphi\rangle\langle\varphi| = \sum_i \sum_j c_i c_j^* |i\rangle\langle j| \qquad (17)$$

where the asterisk denotes a complex conjugate quantity and φ denotes the state function.

By definition the elements of ρ are the "ket times bra" products of the expansion coefficients of the state function in terms of a complete orthonormal base $\{| i > i = 1, 2, \dots n\}$:

$$| \varphi > = \sum_i c_i | i >$$

In a more explicit form, the density matrix elements for our system is written as:

$$\begin{bmatrix} \rho_{11} & \rho_{12} & \rho_{13} & \rho_{14} \\ \rho_{21} & \rho_{22} & \rho_{23} & \rho_{24} \\ \rho_{31} & \rho_{32} & \rho_{33} & \rho_{34} \\ \rho_{41} & \rho_{42} & \rho_{43} & \rho_{44} \end{bmatrix}$$

The diagonal elements ρ_{rr}, ($\rho_{rr} = |\overline{c_r^2}| = P_r$) express the probability that the spin system is found in the eigenstate $| r >$, or, in other words, the population of this state (the horizontal bar denotes the ensemble average, because our microscopic system cannot be viewed as a pure state, characterized by a single state function, but rather as a mixed state, i.e. an ensemble in thermal equilibrium). The off-diagonal elements ρ_{rs}, ($\rho_{rs} = \overline{c_r}\, \overline{c_s}^*$) indicate the coherent superpositions of eigenstates $| r >$ and $| s >$, or coherences. Therefore the transitions between these eigenstates are associated with two coherences, ρ_{rs} and ρ_{sr}.

Any physical measurement can be expressed by a quantum-mechanical operator, Q, whose expectation value, $< \overline{Q} >$, is:

$$< \overline{Q} > = \sum_{r,s} c_r^* \, c_s < r | Q | s > \tag{20}$$

Introducing the density operator previously defined $< \overline{Q} >$ becomes:

$$< \overline{Q} > = \sum_{r,s} < s | \rho | r > \; < r | Q | s > = \mathrm{tr}\{Q \rho\} \tag{21}$$

that is the product of ρ and Q matrices. This means that predictive calculations of the experimental outcome become rather complex. Indeed, even for a simple spin system, the use of the density matrix operator quickly gets cumbersome and, in addition, poorly "descriptive" of the NMR experiment. To overcome these drawbacks, we can resort to the product operators formalism [26]. Provided the spin system is weakly coupled ($\omega_I - \omega_S > 10J_{IS}$), by using this formalism one can follow the fate of all coherences, including those which do not correspond to observable magnetization components, under the influence of rigid rotations describing the evolution due to chemical shift, scalar coupling and radio frequency pulses.

1.3.2 Product Operators

The formalism is based on the products of Cartesian spin operators [26]. The latter are a particular expansion of the density matrix operator in terms of linear combination of products of Cartesian angular momentum operators, I_x, I_y and I_z. The base is completed by 1/2 E (E = unity operator).

The relationship with the density operator is readily appreciated considering the elements of the single spin density matrix, I^α, I^β, I^+ and I^-, where I^α and I^β represent the polarization operators, i.e. the populations of the states α and β ($m^\alpha = +1/2$ $m^\beta = -1/2$), and I^+ and I^- represent the raising and lowering operators, i.e. the coherences, defined as: $I^+ = I_x + iI_y$; $I^- = I_x - iI_y$.
Therefore the Cartesian angular momentum operators are:

$$I_x = \tfrac{1}{2}\left(I^+ + I^-\right)$$

$$I_y = \tfrac{1}{2i}\left(I^+ - I^-\right)$$

$$I_z = \tfrac{1}{2}\left(I^\alpha - I^\beta\right)$$

$$\tfrac{1}{2}E = \tfrac{1}{2}\left(I^\alpha + I^\beta\right)$$

$$\text{(by definition } I^{\alpha/\beta} = {}^1\!/_2\, E\ {}^+\!/\!-\ I_z \text{)} \qquad (22)$$

If the system is formed by two spins I and S, in the density operator we have the new populations

$$I^\alpha S^\alpha,\ I^\alpha S^\beta,\ I^\beta S^\alpha,\ I^\beta S^\beta \qquad (23)$$

the single spin transition coherences

$$I^\alpha S^+,\ I^\alpha S^-,\ I^\beta S^+,\ I^\beta S^-,\ I^+ S^\alpha,\ I^- S^\alpha,\ I^+ S^\beta,\ I^- S^\beta \qquad (24)$$

the two spin transition coherences

$$I^+S^+,\ I^-S^-,\ I^+S^-,\ I^-S^+ \tag{25}$$

i.e. $4 \times 4 = 16$ elements.

Linear combination of these elements give the products of Cartesian angular momentum operators.

For instance:

$$I_x = I_x \cdot 2 \cdot ½ \cdot E = ½ \cdot \left(I^+ + I^-\right) \cdot \left(S^\alpha + S^\beta\right) =$$
$$= ½\left(I^+ S^\alpha + I^- S^\alpha\right) + ½\left(I^+ S^\beta + I^- S^\beta\right) \tag{26}$$

meaning that what we design as transverse x-magnetization of nucleus I is actually a sum of many contributions. The complete list of the Cartesian product operators for a two nuclei spin system includes, as expected, 16 terms:

–	I_x, I_y, S_x, S_y	transverse magnetization components
–	I_z, S_z	longitudinal magnetization components
–	$2I_zS_z$	longitudinal two spin order
–	$2I_xS_z$, $2I_yS_z$	antiphase magnetization of I
–	$2I_zS_x$, $2I_zS_y$	antiphase magnetization of S
–	$2I_xS_x$, $2I_yS_y$ $2I_xS_y$, $2I_yS_x$	two spin coherence
–	½ E	E = unity operator.

A visual representation of transverse and longitudinal magnetization components is given by the classical vector model. A vectorial representation has been recently proposed also for antiphase and two spin coherences [27], as sketched below.

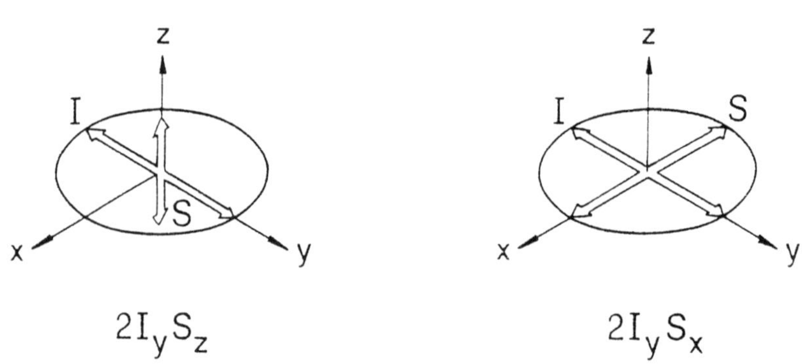

Fig. 10 Vectorial representation of antiphase and two spin coherences. *Reprinted from Ref. 27.*

1.3.2.1 Product Operators Algebra

Any NMR experiment can be described in terms of three types of evolutions, acting on the spin product operators of the system as:

i) rotations due to radio frequency pulses;

ii) rotations due to chemical shift;

iii) rotations due to scalar coupling.

For weakly coupled spin systems the order of application of chemical shift and scalar coupling evolutions, during the free precession intervals, is immaterial, because the corresponding terms in the Hamiltonian commute. A non-selective pulse, acting on spins I and S, can be decomposed in a "cascade" of selective pulses. As far as the result is concerned the time ordering within the cascade is again immaterial. Only the transverse magnetization components, i.e. the (formally) single spin transverse product operators, are observable. The remaining coherences can be observed indirectly, after conversion and/or evolution into transverse magnetization, at the detection stage.

Coherence transfer among different spins can only be brought about by radio frequency pulses.

In order to manipulate product operators, it is worth reminding the convention adopted for rotations and the rigid rotation operator algebra. For spins with positive gyromagnetic ratios, the Larmor frequency vectors are antiparallel with respect to the static magnetic field vector. Thus, if clockwise rotations around the direction of the rotation vector are considered positive, a positive rotation around the z axis leads, when $\gamma > 0$, to the transformation $x \rightarrow y \rightarrow -x \rightarrow -y$. Analogously, the rotation around the x axis leads to $z \rightarrow -y \rightarrow -z \rightarrow y$, while the rotation around y leads to $z \rightarrow x \rightarrow -z \rightarrow -x$. The effect of a rigid rotation on angular momentum operators is calculated by application of an exponential operator. By series expansion of this operator, using the Baker-Campbell-Hausdorff formula and the angular momentum operators commutation rules, one finds, that a rotation of ϑ degrees applied along the y axis to I_z is given by:

$$\exp\left(-i\vartheta I_y\right) I_z \exp\left(i\vartheta\ I_y\right) = I_z \cos\vartheta + I_x \sin\vartheta \quad (27)$$

or, in a more compact form:

$$I_z \xrightarrow{\vartheta I_y} I_z \cos\vartheta + I_x \sin\vartheta \quad (28)$$

The operator ϑI_y is also referred to as propagator.

Keeping in mind these rules we can describe the evolution events taking place during an NMR experiment. The effect of a 90° pulse along the x axis on the longitudinal magnetization component I_z is calculated applying the $90I_x$ operator:

$$I_z \xrightarrow{90_x} = I_z \xrightarrow{90I_x} I_z \cos 90 - I_y \sin 90 = 0 - I_y \qquad (29)$$

Along the same lines, the effect of a non selective $90°_x$ pulse on the y antiphase coherence of I is viewed as a cascade of selective pulses:

$$2I_yS_z \xrightarrow{90I_x} 2I_yS_z \cos 90 + 2I_zS_z \sin 90 \xrightarrow{90S_x}$$

$$\longrightarrow 2I_zS_z \sin 90 \cos 90 - 2I_zS_y \sin^2 90 = -2I_zS_y \qquad (30)$$

The latter transformation illustrates an example of coherence transfer from I to S that can be "graphically" represented as:

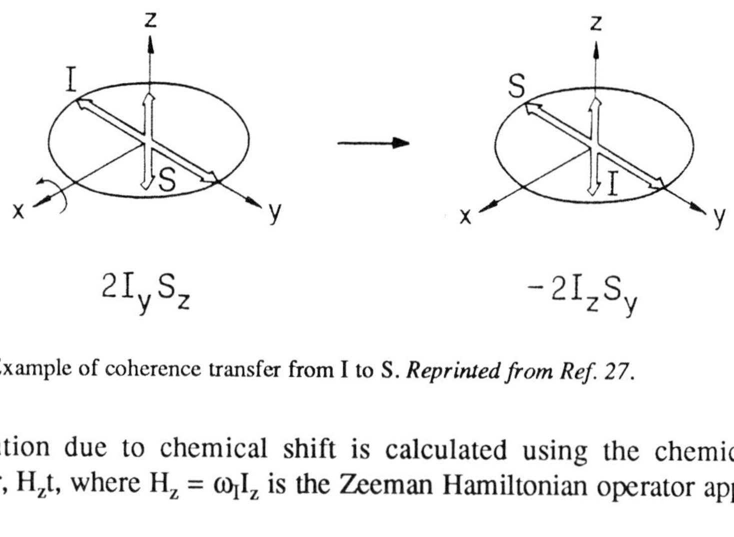

Fig. 11 Example of coherence transfer from I to S. *Reprinted from Ref. 27.*

The evolution due to chemical shift is calculated using the chemical shift propagator, H_zt, where $H_z = \omega_I I_z$ is the Zeeman Hamiltonian operator applied for a time t.
Thus

$$I_x \xrightarrow{\omega_I I_z t} I_x \cos(\omega_I t) + I_y \sin(\omega_I t) \qquad (31)$$

and

$$2I_yS_z \xrightarrow{\omega_I I_z t}\xrightarrow{\omega_S S_z t} 2I_yS_z \cos(\omega_I t) - 2I_xS_z \sin(\omega_I t) \qquad (32)$$

showing that antiphase I coherences are not affected by the S chemical shift propagator (they are invariant with respect to z rotations of S). Finally the scalar coupling evolution is accounted for by the H_Jt propagator, with $H_J = 2\pi J I_z S_z$ scalar coupling Hamiltonian. We can write, for instance:

$$I_x \xrightarrow{2\pi J I_z S_z t} I_x \cos(\pi J t) + 2I_y S_z \sin(\pi J t) \qquad (33)$$

Thus, if t = 1/J

$$I_x \xrightarrow{2\pi J I_z S_z t} -I_x, \text{ i.e. refocusing is achieved,} \qquad (34)$$

while if t = 1/2J

$$I_x \xrightarrow{2\pi J I_z S_z t} 2I_y S_z \qquad (35)$$

the transverse observable magnetization is transformed in antiphase unobservable coherence. Reversing the latter process leads, of course, to transformation of antiphase coherences in observable transverse components.

1.4 The COSY Experiment

By stepping through the COSY pulse sequence, with the insights provided by the product operator formalism, it is now possible to understand where the diagonal and cross-peaks come from. We shall focus our attention on the nucleus I, but the same treatment also applies for the S nucleus.

preparation	evolution	mixing	acquisition
$90°_x$	--- t_1 ---	$90°_x$	t_2

At the end of the preparation period, just prior to the first $90°_x$ pulse, thermal equilibrium is established in the system, with a resulting net longitudinal magnetization component, I_z. The experiment begins with a $90°_x$ non selective pulse, acting on I_z with the propagators $90I_x$ and $90S_x$.

$$I_z \xrightarrow{90I_x} -I_y \quad I_z \xrightarrow{90S_x} -I_y$$

Longitudinal and transverse magnetization components of I are of course unaffected by the S pulse propagator. During the evolution period free precession takes place, under the influence of Zeeman and scalar coupling terms.

$$-I_y \xrightarrow{\omega_I I_z t_1} \begin{cases} -I_y \cos(\omega_I t_1) \\ +I_x \sin(\omega_I t_1) \xrightarrow{2\pi J I_z S_z t_1} \begin{cases} {}^{(1)}\quad I_x \sin(\omega_I t_1)\cos(\pi J t_1) \\ {}^{(2)}\quad 2I_yS_z \sin(\omega_I t_1)\sin(\pi J t_1) \end{cases} \end{cases} \tag{36}$$

From now on the fate of the terms (1) and (2) will be examined separately. The next event in the pulse sequence is the $90°_x$ mixing pulse.
Since term (1) is invariant with respect to any rotation around the x axis, is left unaffected by the pulse and will evolve as:

$$^{(1)} \xrightarrow{90I_x} \xrightarrow{90S_x} I_x \sin(\omega_I t_1)\cos(\pi J t_1) \quad ^{(1a)} \tag{37}$$

The term (2) evolves as:

$$2I_yS_z \sin(\omega_I t_1)\sin(\pi J t_1) \xrightarrow{90I_x} 2I_zS_z \sin(\omega_I t_1)\sin(\pi J t_1) \xrightarrow{90S_x}$$

$$\longrightarrow -2I_zS_y \sin(\omega_I t_1)\sin(\pi J t_1) \quad ^{(2a)} \tag{38}$$

that is, from y antiphase of I it becomes –y antiphase of S. The final stage is the detection period, where again free precession takes place.
For (1a) we have:

$$^{(1a)} \xrightarrow{\omega_I I_z t_2 \omega_S S_z t_2} \begin{cases} I_x \sin(\omega_I t_1)\cos(\pi J\, t_1)\cos(\omega_I t_2) \xrightarrow{2\pi J\, I_z S_z t_2} \quad ^{(1b)} \\ I_y \sin(\omega_I t_1)\cos(\pi J\, t_1)\sin(\omega_I t_2) \xrightarrow{2\pi J\, I_z S_z t_2} \quad ^{(1c)} \end{cases}$$

$$^{(1b)} \quad = I_x \sin(\omega_I t_1)\cos(\pi J t_1)\cos(\omega_I t_2)\cos(\pi J\, t_2) \tag{39}$$

$$^{(1c)} \quad = I_y \sin(\omega_I t_1)\cos(\pi J\, t_1)\sin(\omega_I t_2)\cos(\pi J\, t_2)$$

where (1b) and (1c) represent observable transverse magnetization components of I, exhibiting the ω_I chemical shift both in F_1 and F_2. These components will therefore be observed as diagonal peak, at $\omega_1 = \omega_2 = \omega_I$.
For (2a) we have:

$$(2a) \xrightarrow{\omega_I I_z t_2} -2I_zS_y \sin(\omega_I t_1)\sin(\pi J\, t_1) \xrightarrow{\omega_S S_z t_2} \longrightarrow$$

$$-2I_zS_y \sin(\omega_I t_1)\sin(\pi J\, t_1)\cos(\omega_S t_2) \xrightarrow{2\pi J\, t_2 I_z S_z} (2b)$$

$$(2b) \quad = S_x \sin(\omega_I t_1)\sin(\pi J\, t_1)\cos(\omega_S t_2)\sin(\pi J\, t_2)$$

$$2I_zS_x \sin(\omega_I t_1)\sin(\pi J t_1)\sin(\omega_S t_2) \xrightarrow{2\pi J\, t_2 I_z S_z} (2c) \qquad (40)$$

$$(2c) \quad = S_y \sin(\omega_I t_1)\sin(\pi J\, t_1)\sin(\omega_S t_2)\sin(\pi J\, t_2)$$

Here (2b) and (2c) are the observable transverse magnetization components of S wich evolved under the action of the scalar coupling propagator from antiphase coherence of S. These components exhibit different chemical shift in t_1 and t_2, ω_I and ω_S respectively, and will therefore represent the cross-peaks.

None of the neglected components in this branching tree produce observable magnetization components (they give antiphase, two spin and longitudinal coherences).

Different phase properties of the diagonal and cross-peak can be appreciated by looking at the J modulation factors in t_1 and t_2. The diagonal peak exhibits a $\cos(\pi Jt_1)\cos(\pi Jt_2)$ modulation, whereas the cross-peak has a $\sin(\pi Jt_1)\sin(\pi Jt_2)$ modulation. Their relative phases will therefore differ by 90° in both dimensions.

1.4.1 Longitudinal Relaxation Effects

Relaxation was never taken into account in our discussion. During t_1 evolution, however, some longitudinal relaxation may occur, which would lead to a non zero longitudinal magnetization component at the end of this interval. The mixing pulse would convert it into transverse components without any t_1 labelling. This means that we could observe a copy of the spectrum parallel to F_2 direction, at $F_1 = 0$. These signals are referred to as axial peaks. The artifact can be easily eliminated by increasing the phase of the mixing pulse by 180°. The phase of the axial peaks will be incremented as well, leading to their cancellation in the final spectrum. Using the product operator it can be demonstrated that the 180° phase shift of the mixing pulse does not affect the diagonal nor the cross-peak.

1.5 The Rationale of 2D NMR Strategy

The general strategy adopted for the assignment and the interpretation of 2D NMR spectra of biopolymers entails two steps. The first step assignment is based on the existence of scalar couplings between nuclei connected by two or more chemical bonds. These through-bond interactions are manifested by splitting of the NMR line pattern, as well as by cross-peaks in a 2D map, i.e. information about the chemical shifts of the scalar coupled nuclei. Thus 2D experiments like COSY, multiple quantum filtered COSY, TOCSY, etc., enable the identification of the residue types, according to the specific scalar connectivity patterns associated with their spin systems. When a particular residue occurs only once along the molecule, the identification of its spin system allows the direct assignment of the relative resonances to the specific molecular location. The qualitative spin system identification, however, is not sufficient when a particular residue occurs repeatedly in the sequence.

A second step assignment is required exploiting through-space connectivities. When two or more nuclei are closely spaced (less than 0.5 nm) their mutual dipolar contribution to the relaxation, is reflected by the connectivity map of 2D NOESY and ROESY. The observation of a dipolar connectivity between two nuclei is a direct consequence of a through-space interaction due to the structure of the molecule in solution, and, in this respect, it is clear that there is no upper limit of chemical bond number between dipolarly coupled nuclei. As expected, however, spatial proximity relationships are more likely to occur between nuclei in the same or in consecutive residues. Thus the combined examination of qualitative first step assignments and dipolar connectivities, along with the knowledge of the primary structure, leads to site specific molecular locations. This procedure is better known as sequential assignment. In addition, the observed kind of sequential dipolar connectivity is diagnostic for the local secondary structure, while long range dipolar connectivities (i.e. between nuclei far apart along the primary sequence) provide information about the spatial arrangement of different structural domains.

From these outlines it is readily seen that there is a close relationship between assignment and structural conclusion. Strictly speaking, the whole procedure is based on circular arguments and thus erroneous assignment may be heavily misleading in terms of structural conclusions.

The consistency of the results should therefore be tested very carefully using independent evidence. In the absence of the latter, even the use of sophisticated topology and/or energy minimisation routines may be not sufficient to remove the bias coming from a wrong interpretation of the experimental data.

1.5.1 Size, Shape and Constituents of Biopolymers and NMR Structural Perspectives

The ^1H chemical shift range is a major limitation in NMR studies of biopolymers. The usual proton chemical shift window spans over 10-15 ppm, though it may be as large as hundreds ppm in the presence of contact or pseudo contact interactions due to paramagnetic centres. For instance, in the ^1H 600 MHz spectrum of a 50 residue polypeptide the resonance of 400-500 different protons are concentrated over 6 KHz, and each resonance pattern exhibits a width ranging from 8 to 40 Hz.

Although the shifts induced by secondary and tertiary structure alleviate this overlap to some extent by differentiating the resonance frequencies of each individual spin system type, the molecular dimensions one can cope with using high resolution NMR are still limited and proteins are the most favourable case. Their "building block heterogeneity" is fortunate from the NMR point of view because it gives rise to very characteristic spin system patterns, with relatively wide chemical shift dispersion. Moreover, by virtue of the globular shape of most proteins, their overall tumbling is roughly isotropic and thus their lines broaden very slowly with increasing molecular weight. As a matter of fact, proteins with MW up to 10 KDalton can be studied by 2D NMR, i.e. individual assignment can be obtained for all or nearly all of the approximately one hundred residues of the molecule. Similar figures are absolutely unpractical with polynucleotides and polysaccharides, because of molecular asymmetry, lower variability of the building blocks and narrow resonance dispersion (especially with sugar). Complete assignment can be achieved for molecules containing, at most, 20 base pairs or 8-10 sugar units, roughly. These upper limits of molecular weight are bound to progressively change by virtue of the continuous development in NMR techniques. 3D NMR and heteronuclear indirect detection offer already a viable experimental alternative to overcome the resonance crowding problems and further promising perspectives for both resolution and sensitivity rely on the progress in the superconducting magnet technology.

1.5.2 Appendix

1.5.2.1 Similarity Theorem

$$\Im\{s(at)\} = \frac{1}{|a|} S(\omega/a) = \frac{1}{|a|} S(f/a) \qquad \text{(A.1)}$$

This implies that a broadening of the function in one domain yield a narrowing in the Fourier transformed domain and vice versa.

1.5.2.2 Shift Theorem

This theorem states that if s(t) has the Fourier transform S(ω), for a τ shift of the function along the time axis, the relation holds:

$$\Im\{s(t-\tau)\} = e^{-i\omega\tau} S(\omega) = e^{-i2\pi f\tau} S(f) \qquad \text{(A.2)}$$

It follows that the Fourier transform remains identical except that it gets multiplied by a phase factor which is proportional to the time shift and varies linearly with the frequency. The fundamental importance of this theorem comes from the fact that it describes the reason why differently delayed acquisitions of the signal show different phases in the corresponding frequency responses.

1.5.2.3 Convolution Theorem

The convolution theorem states that if the Fourier transform of s(t) and r(t) are respectively S(ω) and R(ω) then:

$$\Im\{r(t) * s(t)\} = R(\omega) \cdot S(\omega) = R(f) \cdot S(f) \tag{A.3}$$

$$\Im^{-1}\{R(\omega) * S(\omega)\} = \frac{1}{2\pi}\, r(t) \cdot s(t) \tag{A.4}$$

$$\Im^{-1}\{R(f) * S(f)\} = r(t) \cdot s(t) \tag{A.5}$$

where * denotes the "convolution product" which is in general defined as follows:

$$g(t) * f(t) = \int_{-\infty}^{\infty} g(\tau) f(t-\tau) d\tau \tag{A.6}$$

It implies that any filtering process which can be expressed as a convolution, can be transformed into a product in the co-domain.

1.5.2.4 Power Theorem

This theorem is important for sensitivity considerations and states that the signal energy can be calculated by integration in time or in frequency space, as:

$$\int_{-\infty}^{\infty} |s(t)|^2 dt = \frac{1}{2\pi} \int_{-\infty}^{\infty} |S(\omega)|^2 d\omega = \int_{-\infty}^{\infty} |S(f)|^2 df \tag{A.7}$$

References

[1] Purcell, E.H., Torrey, H.C. and Pound, R.V. *Phys. Rev.* (1946) *69*, 37.

[2] Bloch, F., Hansen, W.W. and Packard, M.E. *Phys. Rev.* (1946) *69*, 127.

[3] Ernst, R.R. and Anderson, W.A., *Rev. Sci. Instr.* (1966) *37*, 93.

[4] Jeener, J. Ampere International Summer School, Basko Polje, Yugoslavia, (1971).

[5] Aue, W.P., Bartholdi, E., and Ernst, R.R. *J. Chem. Phys.* (1976) *64*, 2229.

[6] Cooley, J.W. and Tukey, J.W. *Math. Comput.* (1965) *19*, 297.

[7] Bloch, F. *Phys. Rev.* (1946) *70*, 460.

[8] Freeman, R. in "A Handbook of Nuclear Magnetic Resonance", Longman Scientific & Technical Ed., Harlow (UK) p.174 (1987).

[9] Bax, A., *Bull. Magn. Reson.* (1984) *7*, 167.

[10] Bodenhausen, G., Freeman, R., Niedermeyer, R. and Turner, D.L. *J. Magn. Reson.* (1977) *26*, 133.

[11] Ernst, R.R., Bodenhausen, G. and Wokaun, A. "Principles of Nuclear Magnetic Resonance in One and Two Dimensions", Clarendon Press, Oxford (1987).

[12] Aue, W.P., Bachmann, P., Wokaun, A. and Ernst, R.R., *J. Mag. Reson.* (1978) *29*, 523.

[13] Bax, A. and Mareci, T.H. *J. Magn. Reson.* (1983) *53*, 360.

[14] Turner, D.L. *J. Magn. Reson.* (1984) *58*, 462.

[15] "Two-dimensional NMR Spectroscopy", edited by W.R. Croasmun and R.M.K. Carlson, VCH (1987).

[16] Keeler, J. and Neuhaus, D. *J. Magn. Reson.* (1985) *63*, 454.

[17] Bax, A., Mehlkopf, A.F. and Smidt, J. *J. Magn. Reson.* (1980) *40*, 213.

[18] DeMarco, A. and Wüthrich, K. *J. Magn. Reson.* (1976) *24*, 201.

[19] Bax, A., Morris, G.A. and Freeman, R. *J. Magn. Reson.* (1981) *43*, 333.

[20] Campbell, I.D., Dobson, C.M., Williams, R.J.P. and Xavier, A.V. *J. Magn. Reson.* (1973) *11*, 172.

[21] States, D.J., Haberkorn, R.A. and Ruben, D.J. *J. Magn. Reson.* (1982) *48*, 286.

[22] Marion, D. and Wüthrich, K. *Biochem. Biophys. Res. Commun.* (1983) *113*, 967.

[23] Redfield, A.G. and Kunz, S.D. *J. Magn. Reson.* (1975) *19*, 250.

[24] Abraham, A. in "Principles of Nuclear Magnetism", Oxford University Press, Oxford (UK) (1961).

[25] Slichter, C.P. in "Principles of Magnetic Resonance", Springer, Berlin, W. Germany (1978).

[26] Sorensen, O.W., Eich, G.W., Levitt, M.H., Bodenhausen, G. and Ernst, R.R. *Prog. NMR Spectroscopy* (1983) *16*, 163.

[27] Bazzo, R., Boyd, J., Campbell, I.D. and Soffe, N. *J. Magn. Reson.* (1987) *73*, 369.

2. Frontiers in NMR of Paramagnetic Molecules: ^{1}H NOE and Related Experiments

Lucia Banci, Ivano Bertini, Claudio Luchinat and Mario Piccioli

2.1 Introduction

NMR spectroscopy of solutions of paramagnetic molecules is a small part, although well developed, of the large field of NMR [1,2]. The presence of unpaired electron(s) which characterizes a paramagnetic compound gives rise to the so-called isotropic or hyperfine shift [1]. Its extent is related to the unpaired spin density on the resonating nuclei (contact contribution) and to a function of the magnetic susceptibility anisotropy and the unpaired electron(s)-resonating nuclei distance (pseudocontact contribution). The isotropic shifts can be very large, i.e. outside the usual shift range observed for the diamagnetic compounds so that, especially in macromolecules, the signals of the protons sensing the unpaired electron(s) are easily distinguished from the others. Another, perhaps more important, feature of the NMR of paramagnetic compounds is the fast relaxation of nuclei sensing the unpaired electron(s). Both T_1 and T_2 are shortened with respect to the diamagnetic compounds [3]. The contributions to nuclear T_1^{-1} and T_2^{-1} may be again contact [4] and dipolar [5] in nature. They are related to the square of the interaction energy times a function of the correlation time τ_c and the magnetic field. In the case of the contact contribution the correlation time, in absence of chemical exchange, is given by the electronic correlation time τ_s, whereas in the case of dipolar contribution it is given by whichever is shorter between τ_s and the rotational correlation time τ_r. A further contribution to relaxation, dipolar in nature, is the so called Curie relaxation [6,7]. Such contribution is given by the interaction of nuclear spins with the static magnetic moment produced by the difference in population of the electron spin levels due to their Boltzmann distribution. Curie relaxation mechanism does not affect significantly T_1^{-1} while it can strongly affect T_2^{-1}. It is modulated by τ_r and, owing to its dependence on the square of the magnetic field, can provide additional line broadening in the NMR signal.

The short nuclear relaxation times of paramagnetically shifted signals have prevented, until recently, the development of sophisticated techniques like NOE and 2D experiments. The improvement of the instrumentation performances has lead several researchers to further explore this field, owing to the occurence of many paramagnetic species in nature and in the artificial chemical world.

We are going here to review the theory of NOE, some typical examples, and the perspectives with respect to 2D experiments.

2.2 The Theory of NOE

Nuclear Overhauser Effect (NOE) is defined as the fractional change, η_I, in the integrated NMR signal intensity of a nuclear spin I when another spin is saturated, [8,9]

$$\eta_I(t) = \frac{<I_z(t)> - <I_z(\infty)>}{<I_z(\infty)>} \tag{1}$$

where $< I_z(\infty) >$ is the expectation value of I_z for the observed signal in the absence of saturation of the other signal, and $< I_z(t) >$ is the expectation value of I_z when the other signal has been saturated for a time t. The spins involved in this effect can be either heteronuclear or magnetically inequivalent homonuclear spins.

In order to rationalize the extent of the NOE and its relation with the structural parameters of the systems, we refer to two nuclei which are dipolarly coupled [5,10,11]. For simplicity but without loss of generality we can consider two spins I = 1/2, homo- or heteronuclear, I and J. In the coupled system they give rise to four possible eigenstates according to the values of I_z and J_z, as shown in Scheme 1 where + and – refer to the $m_{I,J}$ values of the spins, the first being relative to spin I and the second to spin J.

The system, after a perturbation that has removed the equilibrium condition, will tend again to the equilibrium. This will occur for each spin with a rate which depends on the intrinsic relaxation rate of the spin under consideration, but also on the relaxation rate of the other spin to which it is coupled.

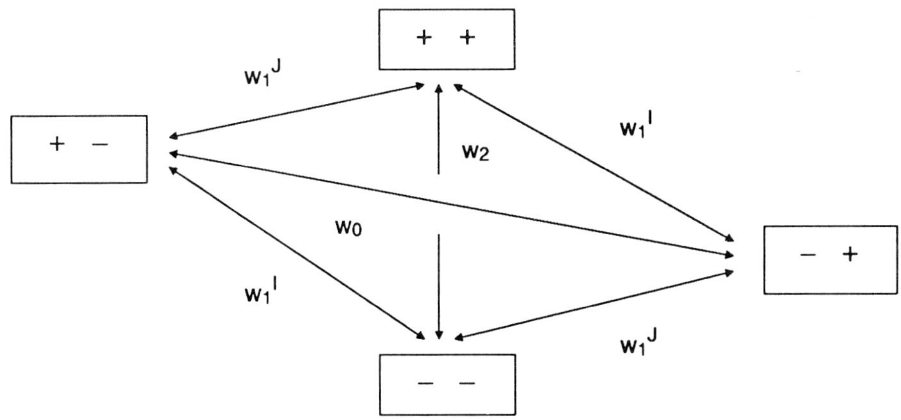

Scheme 1

As it will be shown in the Appendix (Sec. 2.5), the rates of change of the population differences of spin I and J are given by

$$\frac{d\Delta P^I}{dt} = (w_0 + 2w_1^I + w_2)\,(\Delta P^I(\infty) - \Delta P^I) + (w_2 - w_0)\,(\Delta P^J(\infty) - \Delta P^J)$$

$$\frac{d\Delta P^J}{dt} = (w_0 + 2w_1^J + w_2)\,(\Delta P^J(\infty) - \Delta P^J) + (w_2 - w_0)\,(\Delta P^I(\infty) - \Delta P^I) \tag{2}$$

where $w_1{}^I$, $w_1{}^J$, w_0, and w_2 are the transition probabilities between the two involved states as indicated in Scheme 1. $w_1{}^I$ and $w_1{}^J$ are relative to the single quantum transition while w_0 and w_2 are relative to the zero quantum ($\Delta m_I + \Delta m_J = 0$) and the double quantum ($\Delta m_I + \Delta m_J = 2$) transition, respectively.

Following the Solomon formalism [5], it is useful to define

$$\rho_I = w_0 + 2w_1^I + w_2\,; \quad \rho_J = w_0 + 2w_1^J + w_2\,; \quad \sigma_{IJ} = w_2 - w_0 \tag{3}$$

We can now turn to the expectation values of I_z and J_z at time t, $< I_z(t) >$ and $< J_z(t) >$; they are proportional to ΔP^I and ΔP^J, respectively, and therefore, using also the definition of eqn. (3), eqn. (2) can be rewritten as

$$\frac{d\langle I_z(t)\rangle}{dt} = \rho_I(\langle I_z(\infty)\rangle - \langle I_z(t)\rangle) + \sigma_{IJ}(\langle J_z(\infty)\rangle - \langle J_z(t)\rangle)$$

$$\frac{d\langle J_z(t)\rangle}{dt} = \rho_J(\langle J_z(\infty)\rangle - \langle J_z(t)\rangle) + \sigma_{IJ}(\langle I_z(\infty)\rangle - \langle I_z(t)\rangle) \tag{4}$$

ρ_I and ρ_J represent the intrinsic relaxation rate of spin I and J, respectively, in a coupled two-spin system. Their contribution to the overall relaxation rate of I or J depends only on the difference of the value of $\langle I_z \rangle$ or $\langle J_z \rangle$ from their equilibrium value, i.e. on the difference in population of I or J, respectively.

σ is called cross-relaxation rate and its contribution to the relaxation of I depends on the population difference of J.

As it can be seen from eqn. (4), in a coupled two-spin system the approach of a spin to equilibrium does not follow a single exponential behavior; therefore in the presence of cross-relaxation it is not possible to define T_1.

In the case of two dipolarly coupled spins, ρ and σ take the form [5]

$$\rho_{I(J)} = \left[\frac{\mu_0}{4\pi}\right]^2 \frac{2\hbar^2\gamma_I^2\gamma_J^2 J(J+1)}{15\, r_{IJ}^6} \left(\frac{\tau_c}{1+(\omega_I-\omega_J)^2\tau_c^2} + \frac{3\tau_c}{1+\omega_I^2\tau_c^2} + \frac{6\tau_c}{1+(\omega_I+\omega_J)^2\tau_c^2}\right)$$

$$\sigma_{IJ} = \left[\frac{\mu_0}{4\pi}\right]^2 \frac{2\hbar^2\gamma_I^2\gamma_J^2 I(I+1)}{15\, r_{IJ}^6} \left(\frac{6\tau_c}{1+(\omega_I+\omega_J)^2\tau_c^2} - \frac{\tau_c}{1+(\omega_I-\omega_J)^2\tau_c^2}\right) \tag{5}$$

τ_c is the correlation time for the interaction between the two spins I and J; it depends on any mechanism that produces reciprocal reorientation of the two spins. One contribution derives from the motion of the molecule in solution. Another way by which the spins can change reciprocal orientation are internal motions of a group in the molecule [12]. This means that nuclei in the same molecule can have different correlation times.

In any steady state condition the following relationships hold:

$$\frac{d\langle I_z(t)\rangle}{dt} = -\rho_I(\langle I_z(t)\rangle - \langle I_z(\infty)\rangle) - \sigma_{IJ}(\langle J_z(t)\rangle - \langle J_z(\infty)\rangle) = 0$$

$$\frac{d\langle J_z(t)\rangle}{dt} = -\rho_J(\langle J_z(t)\rangle - \langle J_z(\infty)\rangle) - \sigma_{IJ}(\langle I_z(t)\rangle - \langle I_z(\infty)\rangle) = 0 \tag{6}$$

If, let us say, spin J is saturated, then $< J_z(t) >$ will be zero and therefore, when a steady-state condition is reached, i.e. J is saturated for a time long with respect to ρ_I, the first term of eqn. (6) becomes

$$<I_z(t)> = <I_z(\infty)> + (\sigma_{IJ}/\rho_I) <J_z(\infty)> \tag{7}$$

The fractional variation of the integrated intensity of signal I when J is saturated, i.e. the nuclear Overhauser effect, results to be:

$$\eta_I(t) = (<I_z(t)> - <I_z(\infty)>)/<I_z(\infty)> = (\sigma_{IJ}/\rho_I) <J_z(\infty)>/<I_z(\infty)> \tag{8}$$

As $< I_z(\infty) > \propto I(I+1)\gamma_I$ and $< J_z(\infty) > \propto J\,(J+1)\gamma_J$ in the general case of heteronuclear spins it results:

$$\eta_I = (\sigma_{IJ}/\rho_I)\ J(J+1)\ \gamma_J/I(I+1)\ \gamma_I \tag{9}$$

The relative values of J, I, γ_J and γ_I are important for determining the size and the sign of heteronuclear NOE; obviously in homonuclear NOE they cancel each other and eqn. (9) takes the simple form

$$\eta_I = \frac{\sigma_{IJ}}{\rho_I} \tag{10}$$

Substituting the definition of σ_{IJ} and ρ_I (eqn. (3)) we obtain the general equation for NOE in a two-spin system

$$\eta_I = [(w_2 - w_0)/(2w_1^I + w_0 + w_2)\ ] \cdot J(J+1)\ \gamma_J/I(I+1)\ \gamma_I \tag{11}$$

If, in eqn. (11), the expressions for w_1, w_0 and w_2 for dipolar coupling are considered, we can notice that the dependence of η_I on r_{IJ} is cancelled out and therefore the size of NOE is independent on the internuclear distance, depending only on the $\omega_I\tau_c$ value (see later). It is important to know, however, that this simple treatment presumes only dipolar relaxation without other interactions and an isotropic motion. Eqn. (5) shows that η_I is field and τ_c dependent and that it can be both positive or negative; in the fast motion limit, i.e. when $\omega_I\tau_c << 1$, the maximum homonuclear NOE (for two isolated, dipolarly coupled spins) is + 0.5. On the other side, in the slow motion limit, when $\omega_I\tau_c >> 1$, the limit NOE value is – 1.0 (Fig. 1).

In real cases I and J are not isolated but they interact with other spins and therefore other mechanisms than those reported in eqn. (4) will be operative for nuclear relaxation.

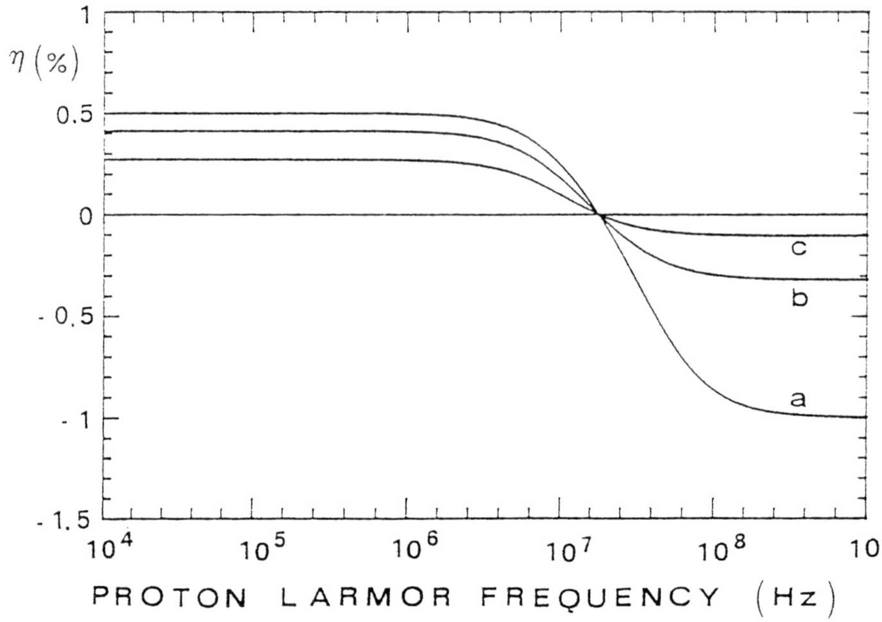

Fig. 1 NOE values as a function of magnetic field, according to eqns. (5) and (12) with τ_c=10 ns, for two hydrogen nuclei 1.8 Å apart. Zero field values of relaxation rates are: a) $\sigma_{I,J}$=11.9 s^{-1}, $\rho_I=\rho_{I(J)}$ (=23.8 s^{-1}); b) $\sigma_{I,J}$=11.9 s^{-1}, $\rho_I=\rho_{I(J)}$ (=23.8 s^{-1})+$\rho_{I(other)}$ (=50 s^{-1}); c) $\sigma_{I,J}$=11.9s^{-1}, $\rho_I=\rho_{I(J)}$ (=23.8s^{-1})+$\rho_{I(other)}$ (=200s^{-1}). $\rho_{I(other)}$ is assumed to be field independent.

In the case of paramagnetic molecules, nuclear relaxation times are dominated by the hyperfine coupling with the unpaired electron(s), being much larger than any other contribution to T_1. In the presence of other contributions to ρ_I ($\rho_{(I)other}$), like in paramagnetic systems, the intrinsic relaxation rate of I increases and therefore the extent of η_I decreases

$$\eta_I = \frac{\sigma_{IJ}}{\rho_I + \rho_{I(other)}} \qquad (12)$$

η_I being smaller the shorter the nuclear T_1.

In such cases the dependence of η_I on the nucleus-nucleus distance is not cancelled out as $\rho_{I(other)}$ is not r_{I-J} dependent. Therefore the presence of a paramagnetic contribution to the intrinsic nuclear relaxation rate makes the NOE magnitude distance dependent even in the extreme case of fast- or slow-motion regimes, and allows the use of this technique, especially in macromolecules, for structural determination.

Note that in any case in a steady state NOE the extent of η_I is independent on the relaxation rate of the irradiated signal as the population of all its spin levels is maintained equal by the application of a selective decoupling radio frequency (as long as constant saturation can be obtained).

In paramagnetic macromolecules the correlation time for $\rho_{I(other)}$ is essentially determined by τ_s and therefore it is independent on τ_r. On the contrary σ_{IJ} is a function of τ_r.

When other mechanisms are operative on nuclear relaxation of spin I, eqn. (12) shows that the NOE values are much smaller than those expected for an isolated two-spin system. Indeed, two protons 1.8 Å apart (a βCH_2 group) with $\tau_c = 10$ ns, give rise to a NOE of -1.0, in the slow motion limit, if the two spins are isolated and only dipolarly coupled ($T_1 = 42$ ms), and a NOE value of -0.1 if the overall T_1 is 4.2 ms (Fig. 1).

If the saturation of J is applied for a time short compared to ρ_I, the system does not reach a steady state condition; in such case integration of eqn. (4) with the condition $J_z(t) = 0$ provides the dependence of the extent of η_I as a function of the saturation time t of signal J.

$$\eta_I(t) = (\sigma_{IJ}/\rho_I)\,(1 - e^{-\rho_I t}) \tag{13}$$

If we use very short saturation times compared with the relaxation time of the nucleus on which NOE is detected ($t \ll \rho_I^{-1}$), this equation becomes:

$$\eta_I(t) = \sigma_{IJ} t \tag{14}$$

This shows that for short saturation times with respect to the intrinsic relaxation time (truncated NOE) the NOE extent is independent on the relaxation time of the nucleus (Fig. 2) and provides a direct measurement of σ_{IJ} [13].

In paramagnetic systems truncated NOE is independent from paramagnetism. Truncated NOE experiments (when possible i.e. when ρ_I is not too short) are useful for obtaining structural information because an analysis of the curve of the build up of NOE as a function of the saturation time provides an accurate determination of the cross-relaxation rate. The analysis of steady-state NOEs (in which saturation is applied for a time long compared to ρ_I) in proteins could be not straightforward, since cross-relaxation connects all the spins in the molecule and a multispin treatment could be necessary. However, with the short saturation times required in paramagnetic systems, the transfer of saturation affects only the nuclei near the saturated one, and 2nd order NOE's do not have time to build. In addition we deal with very small NOE's and 2nd order NOE's, even if they were present, would be too small to be detected. This is one of the great advantages of

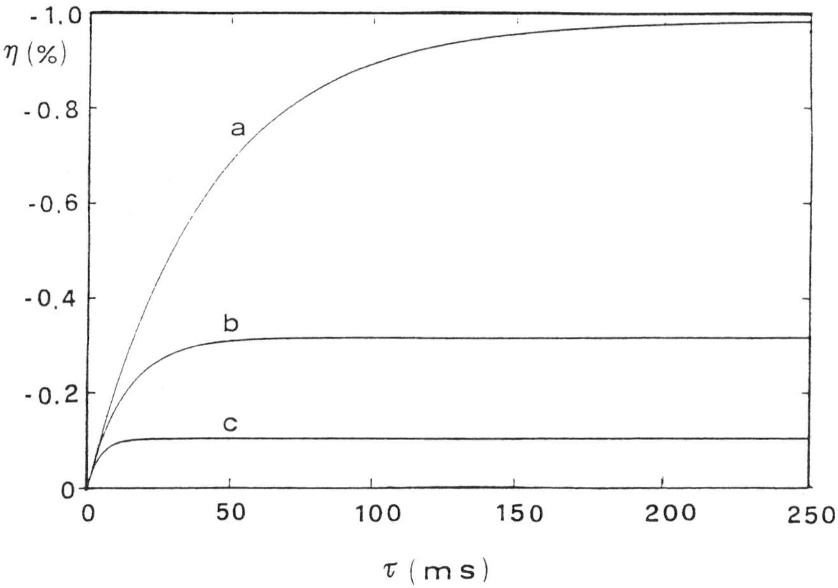

Fig. 2 NOE values as a function of the irradiation time, calculated according to eqns. (5) and (13) at 300 MHz, τ_c=10 ns, for two hydrogen nuclei at 1.8 Å apart. σ_{IJ} and ρ_I values are the same as in Fig. 1.

NOE in paramagnetic molecules. Cross-relaxation rates can therefore be safely analyzed with the two-spin approximation.

In paramagnetic systems we can observe relatively large NOE's in the case of large cross-relaxation rate. This depends, beside the proton-proton distance, on the rotational correlation time τ_r; this shall increase with increasing the molecular weight of the system: the larger the molecule the larger the NOE extent. The use of viscous solvents for increasing τ_r is of help in obtaining larger NOE [14].

The populations of the spin states can be disturbed from their equilibrium condition also by applying strong rf pulses at the resonance frequency of one of the spin. Then the decay of the system to the equilibrium condition is observed. This kind of experiments are called transient NOE [15,16].

In a typical example a selective 180° pulse is applied at the resonance frequency of nucleus, let us say, J, thus inverting its magnetization. Therefore the initial conditions are $I_z(0) = I_z(\infty)$ and $J_z(0) = -J_z(\infty)$. Then, after a delay τ, an observing 90° pulse is applied and the FID collected. The intensity of the signal due to spin I, dipolarly coupled with J, would be changed due to cross-relaxation and a NOE is observed. The building rate is given, in the simplest condition $\rho_I = \rho_J = \rho$, by

$$\frac{d\langle I_z\rangle}{dt} = e^{-(\rho - \sigma)\tau}(1 - e^{-2\sigma\tau}) \tag{15}$$

For any value of σ, i.e. $\rho = -\sigma$ to $\rho >> \sigma$, we can see from eqn. (15) that the rate is faster than in truncated NOE experiments.

Solutions of eqn. (3) with the conditions of this experiment provide the NOE as a function of the delay length.

The overall NOE magnitude as a function of the delay value is reported in Fig. 3. It can be seen that the limits are $\eta_I = 0.385$ for the fast motion limit and -1.0 for the slow motion limit.

As already mentioned, all the previous reasonings on cross-relaxation are developed in the simplest case of two interacting spins. If a more than two center interaction is operative, simple analytic equations for ρ and σ cannot be derived, except in some particular cases [9]. The presence of multicenter interactions produces serious problems in the analysis of the cross-relaxation data. However, as already discussed, this problem almost does not arise in the case of paramagnetic systems.

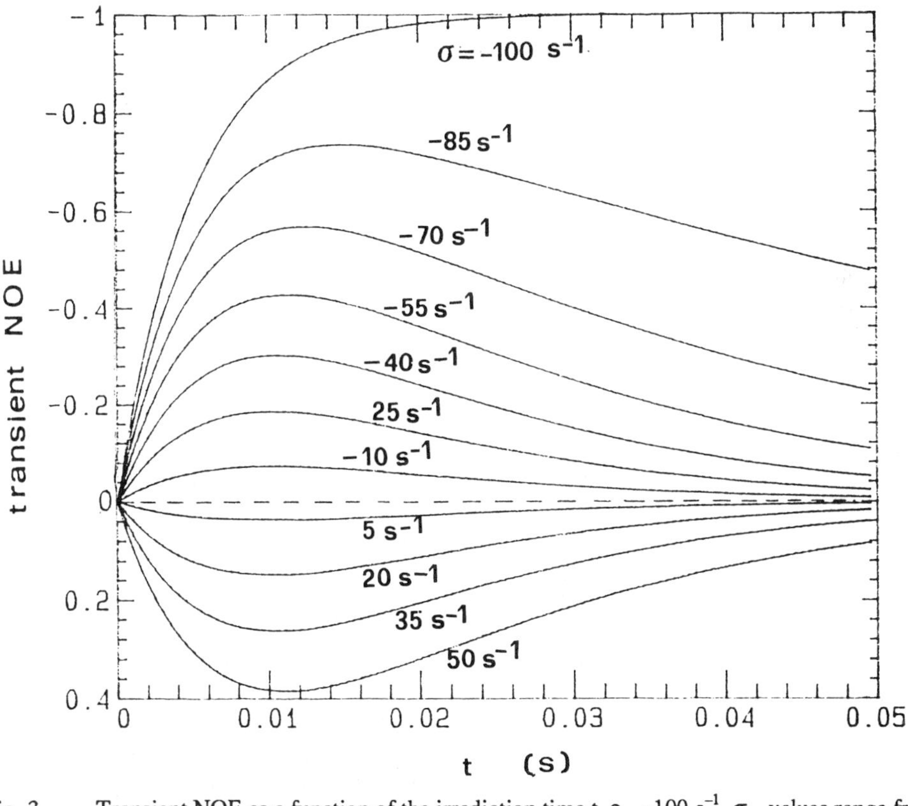

Fig. 3 Transient NOE as a function of the irradiation time t. $\rho_{tot}=100\ s^{-1}$, $\sigma_{I,J}$ values range from $-\rho_{tot}$ (slow motion limit) to $\rho_{tot}/2$ (fast motion limit).

Cross-relaxation can occur also through scalar coupling between the two nuclear spins; in this case only the w_0 transition has a non-null probability and then only a zero quantum transition is effective for scalar cross-relaxation. This transition provides always negative enhancement. However for this mechanism the correlation time cannot be the rotational correlation time but it is determined by the fastest rate between the exchange rate, in the case of chemical exchange, and the nuclear relaxation rate of the nucleus, let us say, J, if nucleus I is observed. This effect would be maximum for spin systems with large coupling constants, small chemical shift differences and exchange rate of nucleus J of the same order than the I-J chemical shift difference. In the case of heteronuclear coupling between a slow relaxing and a fast relaxing nucleus, like quadrupolar nuclei, scalar cross-relaxation could contribute to the overall cross-relaxation of the slow relaxing nucleus.

2.3 Examples of ^{1}H NOE's in Paramagnetic Molecules

The possibility of detecting NOE's in paramagnetic molecules depends on the balance between the T_1^{-1} of the observed proton and the cross-relaxation between the irradiated and the observed proton. The first NOE observed was between the signals of the methyl groups and of the other four non equivalent proton signals of the Ile-99 group in sperm whale metmyoglobin-CN at 360 MHz (Fig. 4) [17,18]. The protein is very soluble, the electronic relaxation time of low spin iron(III) is very short and relatively long are the T_1's of groups not directly coordinated to the metal ion. The magnetic anisotropy is relatively large so that dipolar shifts are operative on protons belonging to groups not directly coordinated to the metal. While assignment of the resonances of the heme ring protons was achieved by using systematic specific deuteriation of all functional groups [19-21], unambiguous assignment of non coordinated residues in the heme pocket is much more difficult and requires the use of more sophisticated methods, like NOE. Signals a, c, d, h, n, x, signal c and d being of intensity three, give NOE among each other plus with a proton in the diamagnetic region. These signals were assigned to the protons of Ile-99. The heme 5-CH_3 signals, which have been assigned through dueteriation techniques, give NOE with signal c; indeed, Ile-99 is close enough to the 5-CH_3 group.

These data indicate that the orientation of Ile-99 relative to the heme is essentially the same in solution and in crystals [18]. Through ^1H NOE experiments, it was possible to show that the heme reconstituted protein is a mixture of two components. The major form has the Ile-99 group dipolarly interacting with the heme pyrrole ring III, while in the minor form it is interacting with pyrrole IV, related to ring III by a 180° rotation about the α-γ meso axis [22].

The NOE technique has been used to assign the CH_3's attached to the same pyrrole ring and between heme CH_3's adjacent to the common meso position in myoglobin reconstituted with hemins methylated at the 2-, 4-, and 6-, 7-, positions and to determine the heme orientation in the holo protein [23]. NOE experiments have allowed the assignment of other signals due to groups not directly coordinated to the iron(III) ion, including the exchangeable protons of the proximal and distal histidines as well as peptide NH's.

As discussed in Section 2 the build up of NOE provides directly σ and, if the interproton distances are known, the correlation time. The latter parameter can be shorter in the case of some H-H pair than another as a result of mobility of the groups containing the observed protons. With this technique it has been shown that the 2-vinyl group in metMb-CN adduct displays significant mobility whereas Ile-99 and βCH_2 of the proximal histidine tumble with the protein [18,24]. Also

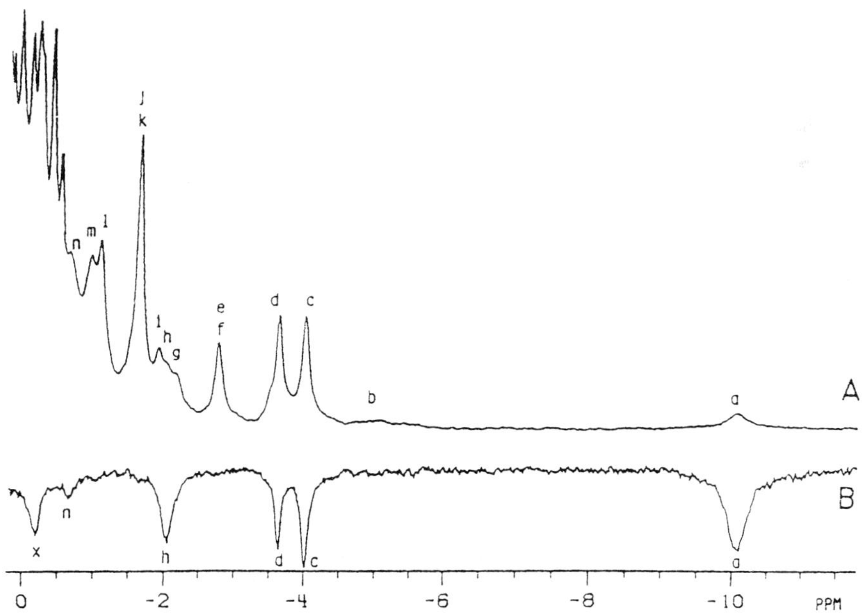

Fig. 4 A) Upfield region of the ^1H NMR spectrum of sperm whale metmyoglobin-CN at 360 MHz; B) steady state NOE difference spectrum obtained upon saturation of peak a. Spectrum intensity is × 5 that of a). *Reprinted from Ref. 18.*

the motional properties of the side chain of Phe CD1 were studied through NOE experiments, thus determining the reorientation rate of the phenyl ring.

Sulphmyoglobin contains a chlorine-like prosthetic group, in which a sulphur atom is incorporated in the hemine skeleton. The assignment of the dicyano sulphhemin complex resonances obtained through NOE allowed the complete description of the electronic structure [25]. NOE experiment on the met-cyano sulphmyoglobin protein has allowed the determination of the chirality of the saturated pyrrole upon sulphur insertion [26].

Horseradish peroxidase-CN (HRP-CN) adduct again contains low spin iron(III); by application of NOE method together with ^{1}H NMR saturation transfer measurements, all 22 heme resonances have been located and assigned, including 12 signals not resolved outside the diamagnetic part of the spectrum [27]. Spin diffusion was observed with irradiation times longer than 100 ms, whereas shorter irradiation times provide only primary NOE. Assignment of the resonances of numerous catalytically important amino acid side chains was also obtained. Residues on the proximal side (His-170, Leu-237 and Tyr-185) and on the distal side (Arg-38, His-42 and Phe-41) were shown to have structural arrangements similar to the same residues in cytochrome c̲ peroxidase, for which the X-ray structure is known [28]. The assignment of signals of exchangeable protons of the distal and the proximal histidines permitted the determination of several crucial aspects of the structure of the heme cavity and the proposal of model for HRP compound II formation. Also in this case time dependent NOE's have been used for the investigation of the mobility of groups inside the heme pocket and trans and cis conformations have been revealed for 2-vinyl and 4-vinyl groups, respectively [27].

NOE experiments on the cyanide adduct of cytochrome c̲ peroxidase from yeast have allowed the complete assignment of all the hyperfine shifted signals [29]. In addition signals from residues not directly bound to the metal ion have been assigned and the relative orientation of the heme pyrrole II substituent have been determined. NMR characterization, including NOE experiments, has shown the close similarity of the structural arrangement around the heme ring in the ferricytochrome c̲ from horse and tuna [30], while the isoenzyme-2 from yeast presents the heme environment similar but not identical to that in the horse and tuna ones [31].

Recently the NOE has been applied successfully also to other than heme proteins. Among non heme iron proteins, NOE has been observed in reduced Fe_2S_2 ferredoxins where it is present an iron(II) and an iron(III) ion magnetically coupled through two inorganic sulphurs that bridge the two metal ions [32]. The T_1 of the signals, especially of the protons sensing iron(III), are quite short providing very small NOE values. Nevertheless connections between the βCH2 and the α CH signals of the protons of the iron (III) site and among the βCH2 signals of the iron(II) site have been observed allowing the assignment of the NMR spectrum and the correlation between the redox potentials of the iron ions with their environment and the interactions with the groups nearby [33].

Obviously NOE can be detected in systems with longer T_1's. In the case of oxidized Fe_4S_4 ferredoxin from *Clostridium Pasterianum* two iron(II) and two iron(III) are present with a magnetic susceptibility value corresponding to one unpaired electron at room temperature. The T_1 of the βCH_2 protons of cysteines range from 4.5 to 13.5 ms. NOE's as large as 5% are observed for six of the eight βCH_2 groups [34].

NOE's have been detected between the βCH_2 protons of an aspartate residue coordinated to tetrahedral high spin cobalt(II) systems [35]. The protein is superoxide dismutase (SOD) with MW 32,000 whose active site contains a copper and a zinc ion bridged through an imidazolato group of a histidine. Zinc can be easily substituted, without loss of activity and structural variation, with cobalt(II) ion [36]. The derivative containing only the cobalt(II) ion in the zinc site shows a spectrum with signals characterized by very short T_1 (1-2 ms). Nevertheless sizeable NOE has been observed which has been used for assignment.

In Cu_2Co_2SOD the copper ion is coordinated to four histidines, one of which is that bridging the two metal ions [36]. The bridge gives rise to magnetic exchange coupling between the two metal ions and, as a result of it, the copper(II) metal ion has now short electronic relaxation times [37]. The T_1 of His protons of the copper domain are about 2-5 ms. NOE's have been observed between vicinal protons of the same histidine ring (Scheme 2).

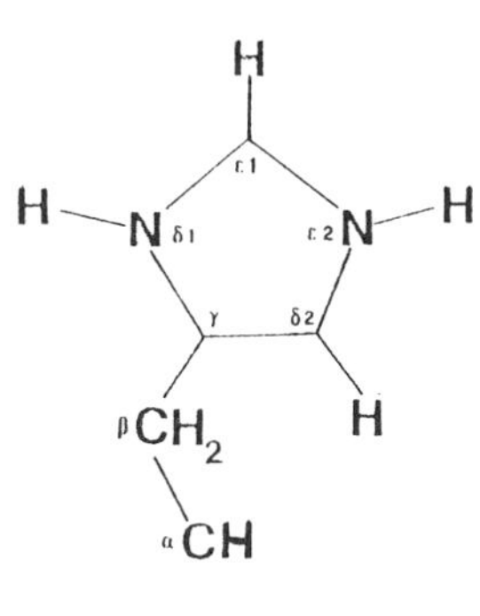

Scheme 2

In the case of His binding through Nδ1, saturation of the NH signal provides only one NOE among the paramagnetic signals, whereas in the case of Nε2 binding to the metal two NOE's are expected. So, it is possible to distinguish between the two types of coordination as depicted in the above scheme. NOE between inter histidine ring protons have been observed up to a distance of 3.2 Å; NOE values smaller than 1% in H_2O solutions and 0.5% in D_2O solutions have been detected (Fig. 5) [38].

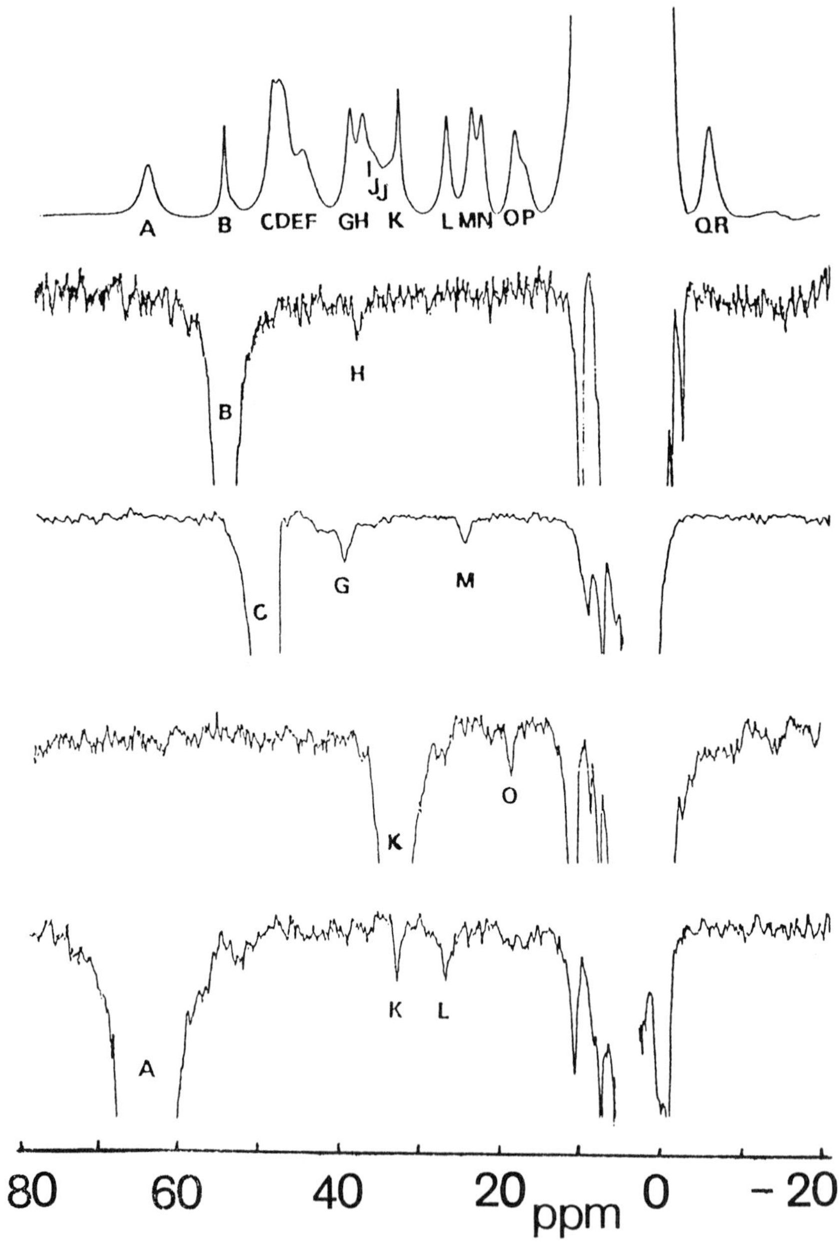

Fig. 5 a) ^{1}H NMR spectrum of Cu_2Co_2SOD in H_2O at 200 MHz and some representative NOE difference spectra which have been crucial for the assignment of the ^{1}H NMR spectrum. *Adapted from Ref. 38.*

The general strategy is to find a distance between two protons through eqn. (5) and then to assign them through comparison with structural data. Therefore NOE's become a very powerful tool for assigning signals in paramagnetic molecules, especially proteins. In the case of Cu_2Co_2SOD an extensive NOE investigation allowed us the assignment of all the paramagnetically shifted resonances (Fig. 6). It is possible that the number of such cases will increase in the near future and that after NOE's have been observed between the hyperfine shifted signals and some diamagnetic signals, then the structure solution can proceed for the diamagnetic part with the more classical 2D and 3D NMR techniques.

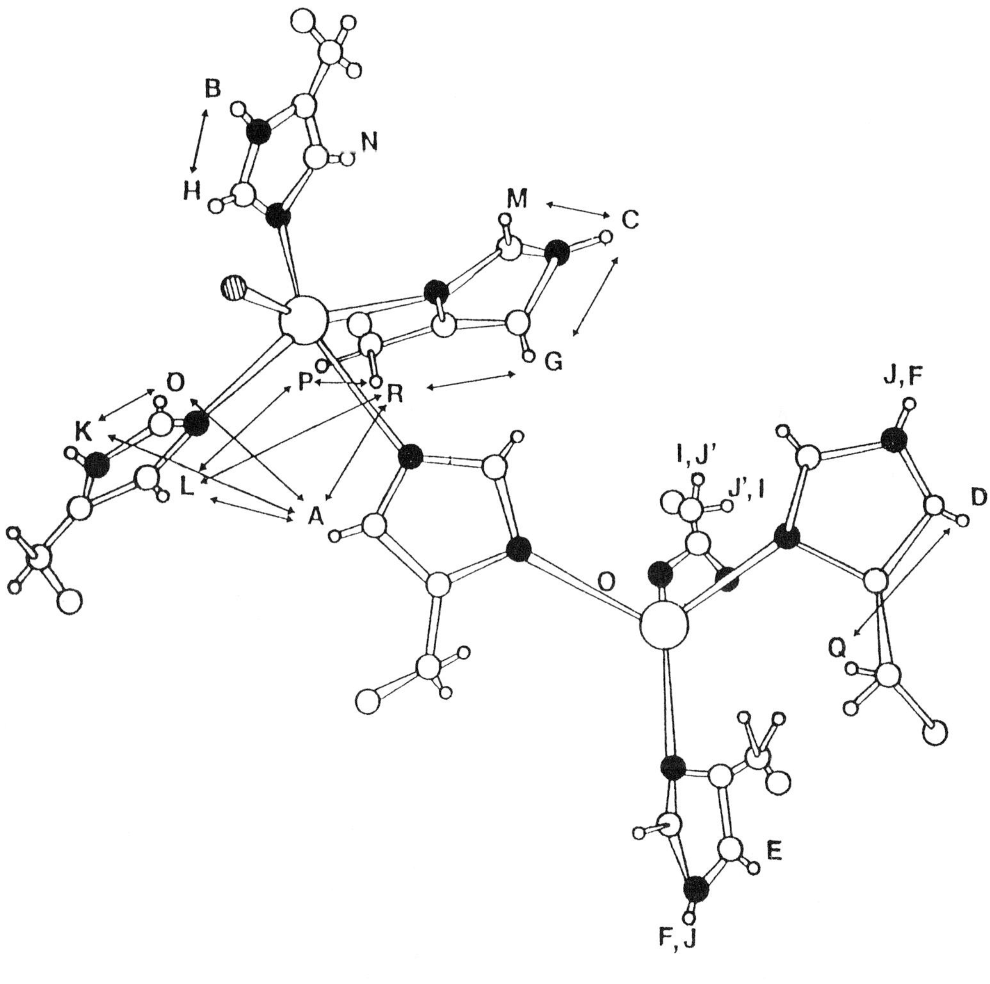

Fig. 6 Schematic drawing of the active site of bovine Cu_2Co_2SOD [40] showing the correlations obtained through ^1H NOE and the assignment of the signals.

2.4 Non-scalar Magnetization Transfer in 2D Experiments

2.4.1 NOESY and EXSY Experiments

2D spectroscopies have obvious advantages with respect to one-dimensional experiments [39,40]. Whereas the scalar magnetization transfer has been treated in Chapter 1, we treat here the dipolar coupling and chemical exchange, devoting a few pages to a pictorial description of the phenomena. This part can be also useful for the following chapters. The potentiality of the technique with respect to paramagnetic systems is then discussed.

Dipolar interactions may be revealed using the two-dimensional NOE experiment NOESY [41-44]. This experiment is one of the few 2D experiments that can be understood using the classical vector representation of the spin magnetization in the rotating frame. As any 2D experiment it can be described in terms of a sequence of four time intervals called preparation, evolution, mixing, and detection periods. The evolution period is called t_1, and the detection period t_2. The latter is the acquisition time in a monodimensional experiment. The sequence of events is illustrated in Fig. 7. The preparation period is a $90°_x$ pulse; after a variable time t_1, a second $90°_x$ pulse is applied followed by a fixed mixing period t_m. Finally, a third $90°_x$ pulse is applied and the FID recorded during the time t_2.

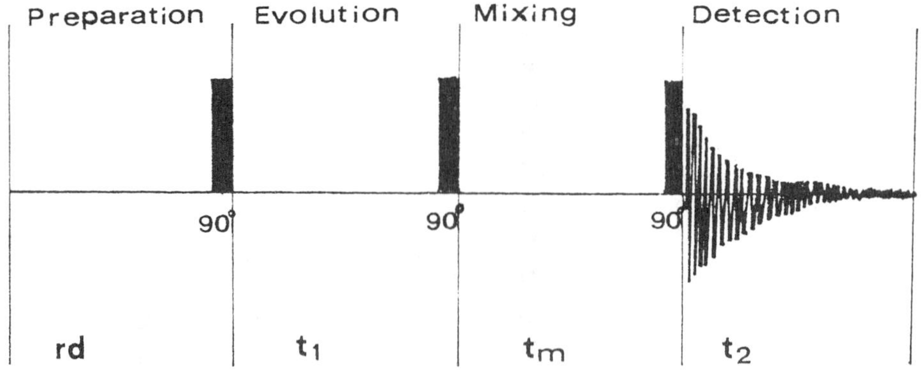

Fig. 7 Sequence of events in NOESY according to the usual representation of a 2D experiment; magnetization transfer occurs during the mixing time t_m.

In 2D experiments a series of FID's are collected at increasing t_1 values. The number of FID's, N_1, will determine the resolution in the resulting ω_1 dimension, just like the data points in a FID determines the resolution in a monodimensional spectrum and in the ω_2 dimension of a 2D experiment, whereas the spectral width in the ω_1 dimension is inversely proportional to the time increment in the t_1 dimension. This time increment corresponds to the dwell time in monodimensional spectroscopy, i.e. the time interval between two adjacent points in the FID. A two dimensional spectrum can be obtained by first Fourier-transforming each of the N_1 FID's into the corresponding spectrum; then, the data can be arranged in such a way as to group the first data point in each spectrum, in the order of increasing t_1 value, in a FID-like function, called interferogram. The same is done with all the data points in the spectra. The resulting N_2 interferograms (if N_2 is the number of data points in the t_2 dimension) are then Fourier-transformed and the resulting 2D spectrum displayed in matrix form.

Let us follow what happens to the magnetization in a system constituted by two dipolarly coupled $I = 1/2$ spins during a NOESY experiment and how the original FID's, the interferograms obtained after the first Fourier transformation, and the final 2D spectrum look like. The first $90°_x$ pulse tilts the magnetization of both spins from the z to the y axis (Fig. 8). Let us assume that the carrier frequency is lower than both Larmor frequencies and that the rotating reference frame is rotating at the carrier frequency. During the time t_1 the two magnetization vectors A and B will slowly precede in the xy plane, the slower (B) being that associated with the spin having its Larmor frequency closer to the carrier. If $t_1 << \Delta\,\omega_A^{-1}$,

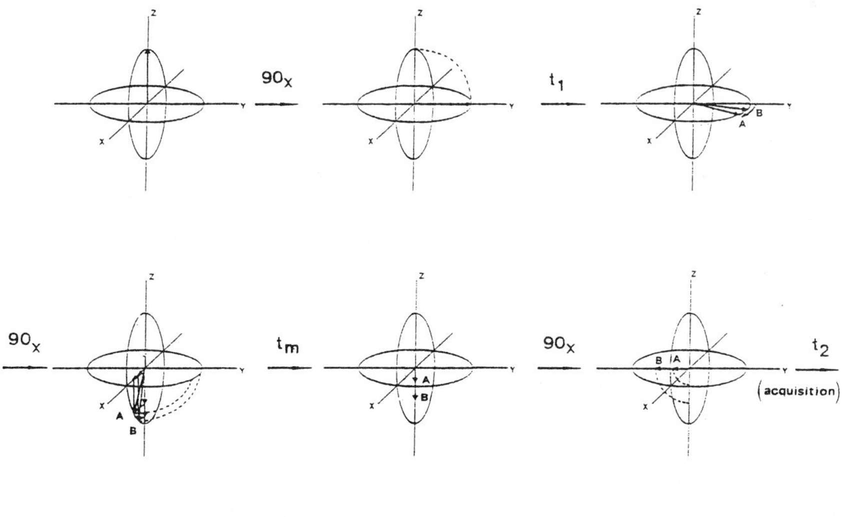

Fig. 8 The magnetization vectors of nuclei A and B during a NOESY pulse sequence: chemical shifts evolve during t_1 and t_2; magnetization transfer occurs during t_m.

$\Delta\omega_B^{-1}$, where $\Delta\omega$ is the shift of each signal from the carrier in rad s^{-1}, the two vectors have preceded little in the first quadrant, and the second $90°_x$ pulse brings them in the –zx plane, with B closer than A to the –z axis. Therefore there will be components of A and B magnetization both along x and –z. Only the latter will be detected by the last $90°_x$, and we will neglect the components along x from now on. The components along –z will evolve during the mixing time t_m according to their longitudinal relaxation times. This is the central step in the experiment: during this time cross-relaxation is operative if the two spins are dipolarly coupled, and the two magnetization vectors will not evolve to their equilibrium values according to single exponentials, but rather with a biexponential behavior. For $t_1 \cong 0$ characteristic of the first experiment in the 2D array the two vectors are essentially fully tilted at –z by the second pulse and the recovery would be that expected in the so-called non-selective T_1 experiment. Fourier transformation will give the spectrum shown at the bottom of Fig. 9, where the intensities of the two signals reflect the state of the magnetization after the time t_m. In the second experiment with t_1 = 100 μs the two vectors will have preceded further in the first quadrant and their components in the –z direction after the second pulse will be i) smaller and ii) more different one from the other. This means that cross-relaxation effects during t_m will be established earlier because of the initial difference in magnetization between the two spins. The situation starts being closer to a selective T_1 experiment, where the two starting magnetizations are made very different

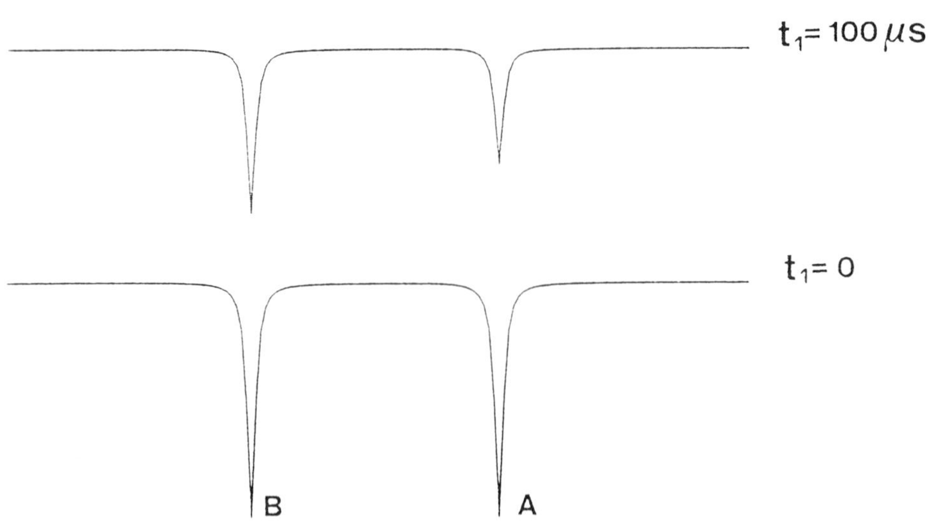

Fig. 9 Evolution of spectra in the ω_2 dimension as a function of t_1 for two peaks with the same line width. At t_1=0 signal A and B are not discriminated by t_1 evolution and possess the same intensity. At t_1=100 μs signal A, which is the most shifted one, is smaller in intensity than B (see Fig. 8).

by selectively inverting only one of them. The transformed signals are shown at the top of Fig. 9. With increasing t_1 values, the two spins will be more and more dephased, up to a phase difference of 180° in the xy plane. In this situation the second pulse will leave a positive z component for one signal and a negative component for the other. A limit case would be that of one magnetization being along y and the other along –y at the time of the second pulse: this results in a full magnetization along z for one signal and a full magnetization along –z for the other, i.e. just what happens in a selective T_1 experiment. Here the effect of cross-relaxation is maximal at the beginning of t_m; this is reflected in the maximum effect on the relative intensities of the two signals after the Fourier transformation.

When all the FID's have been collected and Fourier transformed in the ω_2 dimension, the array of spectra will look like that of Fig. 10. The interferograms are obtained by collecting corresponding points along the t_1 dimension. Of course, where there are no signals the interferogram is flat and so will be its Fourier transform. The interferograms corresponding to each of the two peaks, indicated by dashed lines in Fig. 10, show the typical behavior of an FID. They result from sinusoidal variations in the amplitude of the two signals. These variations arise from the fact that the vectors precessing in the xy plane are sampled by their projection on the z axis, the latter being obtained by tilting them into the xz plane. When a vector is along y its projection along z is negative and largest in absolute value, when it is along x its projection is zero, and when it is along –y its projection is positive and largest. Therefore, the intensity of the signal, say signal

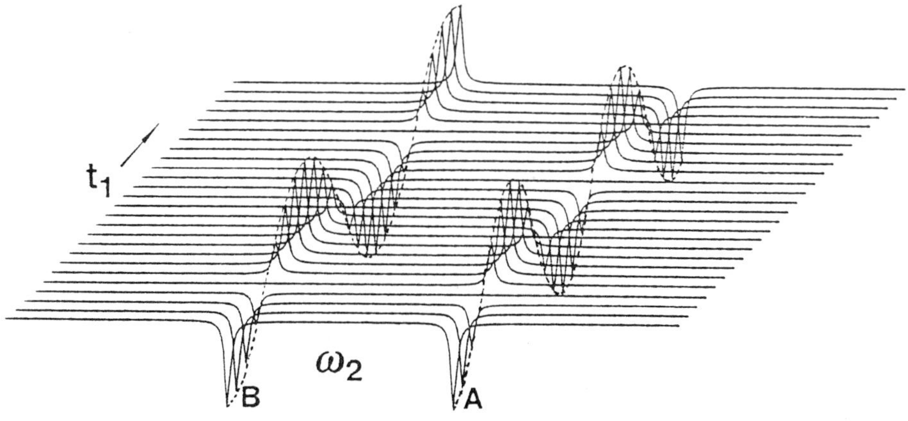

Fig. 10 A 2D data set Fourier transformed in the ω_2 dimension. The data consist of spectra of the type of those shown in Fig. 9. The dashed lines show the modulation in intensity of the two signals and define the interferograms which originate the ω_1 dimension of the 2D spectrum after the second Fourier transform.

A, along the t_1 axis is mainly modulated by the Larmor frequency of the signal itself, ω_A. Fourier transformation along the ω_1 dimension will surely give a strong signal at ω_A just like in the ω_2 dimension, i.e. a diagonal peak. The same will happen for signal B. However, we have seen that during t_m cross-relaxation is operative, and therefore the intensities of the two signals will not be exactly what expected if they were isolated. In turn, cross-relaxation effects are modulated by the fact that the two signals may end up onto the z axis with similar or very different magnetization values depending on their being in phase, antiphase or intermediate situations in the xy plane. This further modulation of the intensity of one signal, say A, therefore occurs at a frequency $\omega_B - \omega_A$; the latter combines with the main oscillation at ω_A to give a frequency ω_B. Since the change in amplitude is small there will be a small but observable signal at ω_B in the ω_1 dimension resulting from Fourier transformation of the interferogram of signal A. For the same reason, there will be a small signal at ω_A in the ω_1 dimension from the interferogram relative to signal B. Therefore the 2D spectrum will look like the one shown in Fig. 11.

The acronym EXSY stands for EXchange SpectroscopY. Chemical exchange is, in fact, an obvious way of transferring magnetization from one spin system to another because a nucleus carries its spin state with it when it jumps from one chemical environment to another. If chemical exchange is slow with respect to the separation in chemical shifts, but still of the order of the relaxation times, some magnetization transfer can take place before the system has relaxed back to equilibrium.

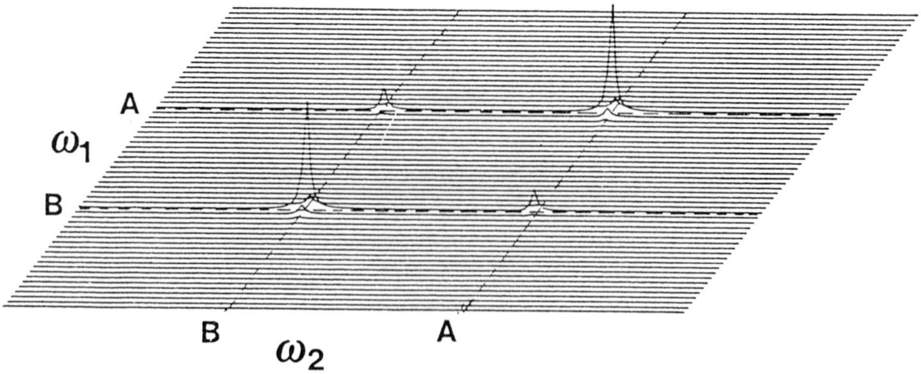

Fig. 11 Stacked plot of a 2D NMR spectrum. Diagonal peaks are from lower left to upper right, cross-peaks indicate the presence of NOE between diagonal peaks.

The steady state magnetization transfer, X, is defined, in analogy with the nuclear Overhauser effect, η, in eqns. (1) and (12), as

$$X = \frac{\langle I_z(t)\rangle - \langle I_z(\infty)\rangle}{\langle I_z(\infty)\rangle} = -\frac{k_{AB}}{k_{AB} + \rho_{tot}} \tag{16}$$

where k_{AB} is the chemical exchange rate constant (assuming a one-step process) between the A and B sites and ρ_{tot} is the overall nuclear relaxation rate of the signal on which the effect is being observed.

It is apparent that, when $k_{AB} > \rho_{tot}$, X approaches unity, i.e. magnetization transfer is most effective. In the NOE experiment this would correspond to $\sigma_{IJ} = -\rho_I$ with $\rho_{I(other)} = 0$. Since the physical mechanism for magnetization transfer is analogous, the same pulse sequence used for NOESY gives an EXSY response in the presence of chemical exchange. The 2D spectrum shows diagonal and cross-peaks of the same sign, as in NOESY spectra in the slow-rotation regime, i.e. when σ_{IJ} is negative.

2.4.2 Perspectives in Paramagnetic Systems

As we have outlined before, the major drawback in performing 2D experiments in paramagnetic systems is the fast nuclear relaxation rates [1]. Ideally, the mixing time t_m should be long enough to allow the magnetization transfer to take place, but short enough not to cause signal loss due to relaxation. The magnetization transfer rate in NOESY is proportional to σ_{IJ}, which can be at most equal to $-\rho_I$. Therefore, the large $\rho_{I(other)}$ operative in paramagnetic systems causes the intensity of the cross- peaks to be always very small. In EXSY the situation can be more favourable since, no matter how large ρ_{tot} is, it may happen that k_{AB} is even larger. In such cases, strong cross-peaks can be present even under fast relaxation conditions.

EXSY data on paramagnetic systems have started to appear, along with the overcoming of instrumental difficulties which are to some extent in common with 2D experiments on diamagnetic molecules. As an example, we report in Fig. 12 the EXSY spectrum of the Pr(III) complex of diethylenetriaminepentaacetate (DTPA) which allowed to address the structural and dynamic aspects of this complex in solution [45]. Chemical exchange between two enantiomers has been revealed, the eight-coordination of the complex and a water molecule as ninth ligand has been focused. The present limit in our capability of observing EXSY

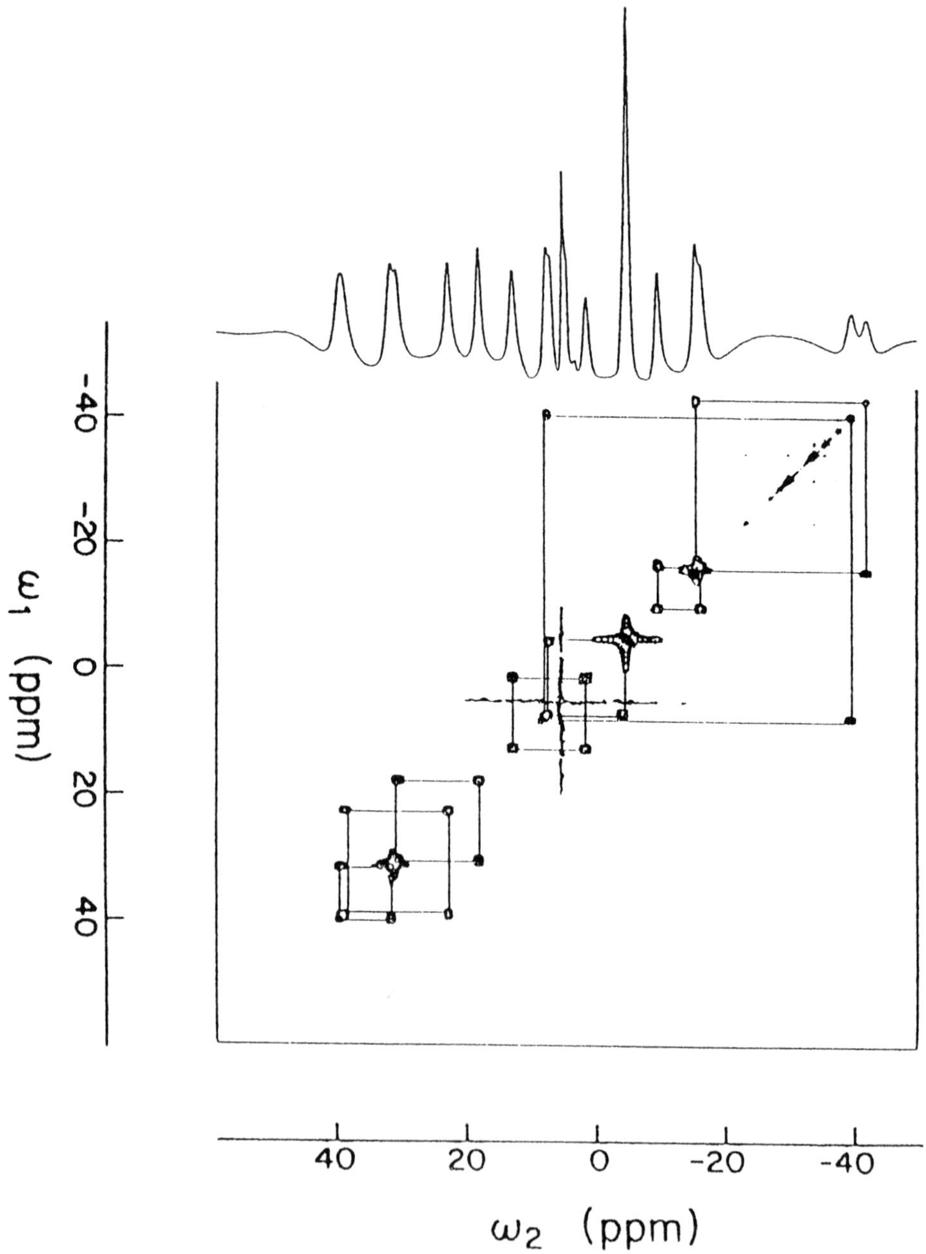

Fig. 12 Contour plot of the EXSY spectrum of 0.3M $Pr(DTPA)^{2-}$ at 500 MHz and 25°. Mixing time was 15 ms. The data was processed using a Lorentzian multiplication corresponding to 10 Hz line broadening in F2 and a convolution difference in F1. The spectrum has been simmetrized about the diagonal. *Reprinted from Ref. 45.*

peaks is reached in compound 5-Cl-NiSal-MeDPT, where cross-peaks arising from the exchange of two non equivalent sides of a chiral chelating agent could be detected even for proton pairs with T_1's as short as 10 ms [46].

Fig. 13 shows the intensity of both diagonal and cross- peaks as a function of t_m for several exchange rate k values and for ρ_{tot} of 100 s^{-1}. Cross peaks have the maximum for

$$t_m = (2k)^{-1} \log(1 + 2k/\rho_{tot}) \tag{17}$$

if $k/\rho_{tot} < 1$ the maximum cross-peak intensity is for $t_m = 1/\rho_{tot}$.

The ratio between cross-peaks and diagonal peaks reaches significant values only for long t_m, i.e. when the intensity of both diagonal and cross-peaks is too low to give rise to a 2D spectrum. If the exchange rate is of the order of magnitude of overall relaxation rate the ratio between cross-peaks and diagonal peaks is about 0.5 when cross-peaks have the maximum intensity, and is about 1 when the intensity of diagonal peaks is around 10% of that at $t_m = 0$.

Fig. 14 shows the EXSY spectrum of 5-Cl-NiSal-MeDPT recorded with t_m of 5 ms.

Chemical exchange can also be detected performing 2D rotating frame experiments, in which the magnetization vector M_z is spin-locked to the carrier frequency [47]. Theoretical background of these experiments has been extensively studied by Ernst and co-workers [39,48,49]. After a 90° pulse, a spin locking pulse sequence is applied during mixing time and FID's are then collected. Both scalar and dipolar coupling can be detected in the rotating frame. By the application of a continuous spin-lock field a nuclear Overhauser enhancement is detected (Rotating frame Overhauser Enhancement SpectroscopY, ROESY) [47,50], while the application of an Homonuclear Hartmann-Hahn [51] pulse sequence (HOHAHA) as the widely used MLEV-17 [52,53] allows detection of all scalar connectivities (TOtal Correlation SpectroscopY, TOCSY) [49].

Application of rotating frame techniques to paramagnetic molecules seems to be feasible not only for small paramagnetic complexes but also for the investigation of paramagnetic macromolecules of biological interest.

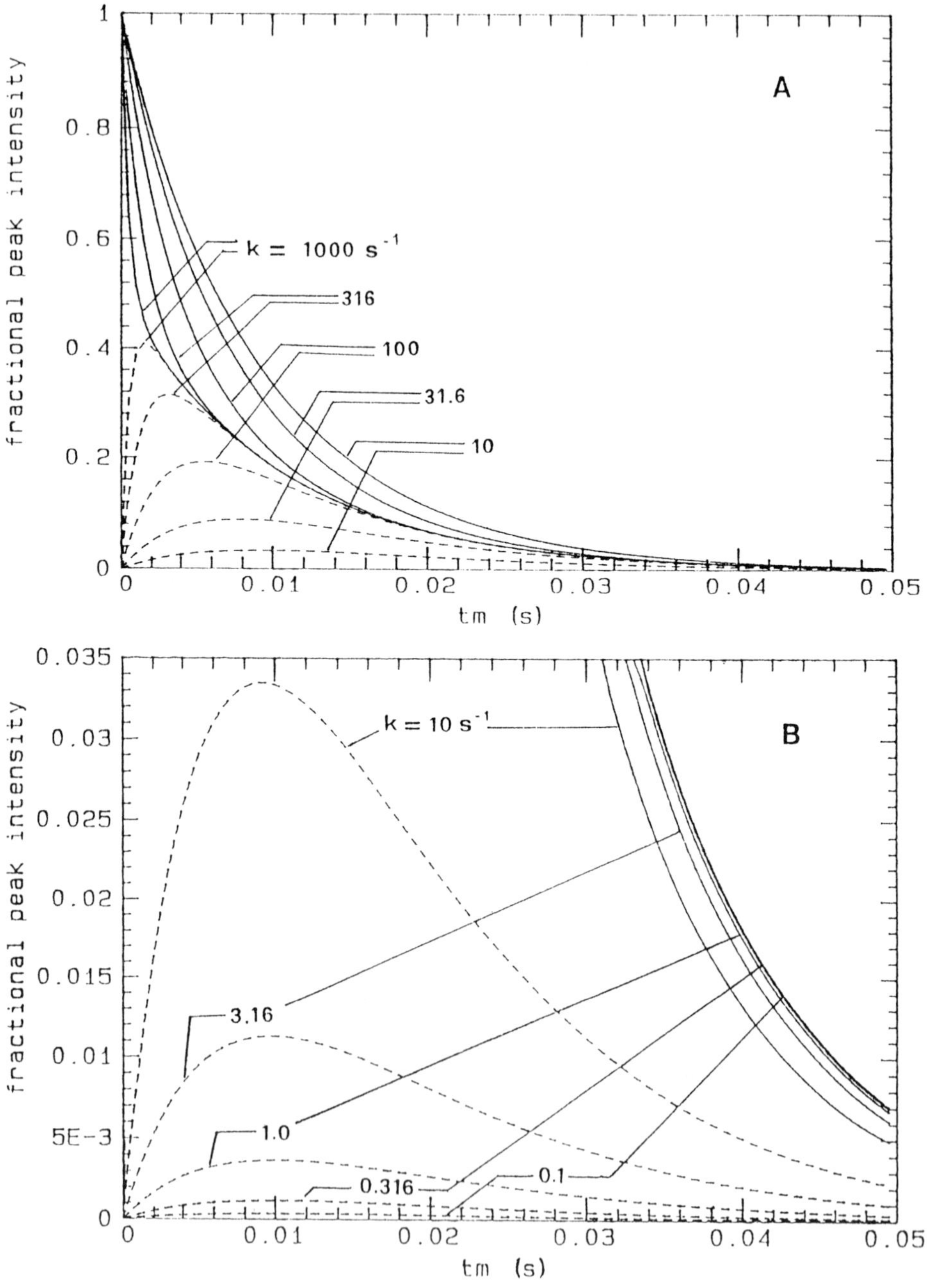

Fig. 13 Diagonal peaks (——) and cross-peaks (----) as a function of mixing time t_m, calculated for ρ_{tot}=100 s^{-1} and several k values. In part A peak intensities for k/ρ_{tot} ranging from 10 to 0.1 are reported; part B show curves calculated for lower k values.

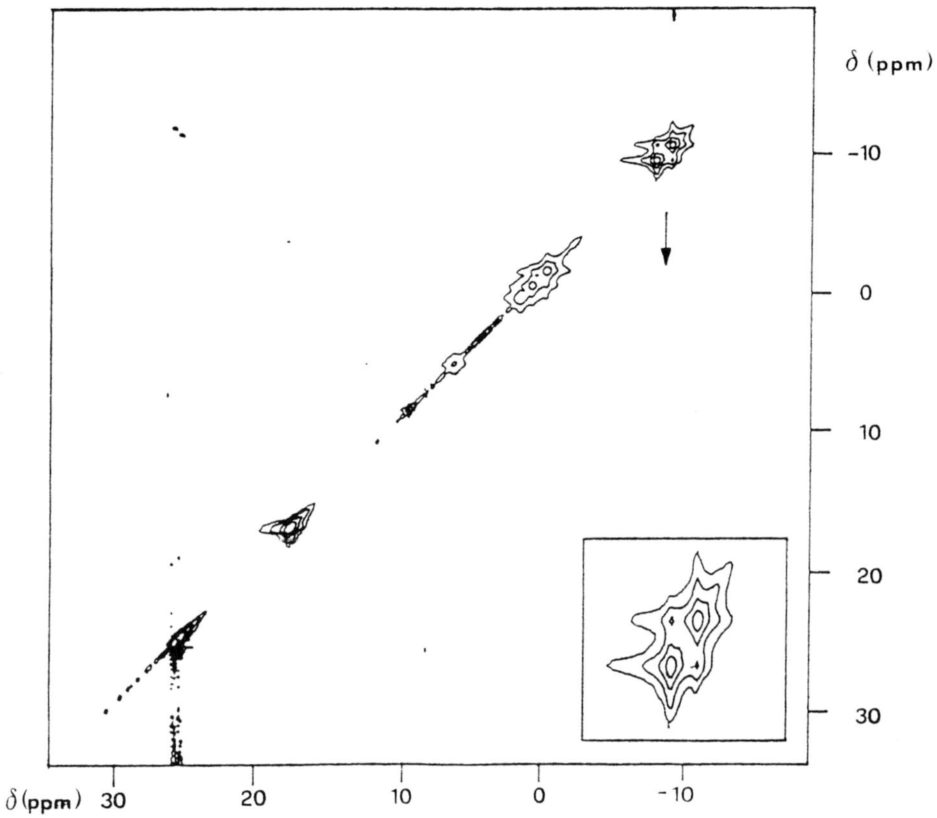

Fig. 14 EXSY spectrum of 5-Cl-NiSal-MeDPT, recorded at 300 MHz. 1K × 2K matrix data point was collected. Sample was dissolved in $CDCl_3$. Mixing time of 5 ms was used. Inset shows the EXSY peaks between the 3H ring proton signals. *Reprinted from Ref. 46.*

2.5 Appendix

Let us define our system as constituted by two spins I and J both with quantum number 1/2. They may be alike (i.e. they constitute a single ensemble) or unlike; in the latter case we are interested in the relaxation behavior of, say, spin I, with J belonging to the lattice. In the coupled system I-J there are four possible eigenstates according to the values of I_z and J_z as shown in Scheme 1 of the text.

In that scheme the + – and – + levels (+ = 1/2; – = –1/2) are at the same energy in the case of alike spins. If the system is perturbed, for example with the appropriate radio or micro frequencies, a population distribution different from that at equilibrium is obtained. After switching the rf power off, the system will tend to equilibrium. We define N_{++}, N_{+-}, N_{-+}, and N_{--} as the populations of the levels at any time and w_1^I, w_1^J, w_0, and w_2 as the transition probabilities, i.e. the probability for the transition between the two involved levels to occur per unit time (with unit time shorter than T_1) per unit spin.

From Scheme 1, the rates of changes in population towards the equilibrium (beside a constant that is proportional to the population difference at equilibrium, according to the Boltzmann distribution) are given by the sum of the rates for the individual transitions in both directions, each rate for a m → n levels transition being the product of w_{mn} times the population of the mth level.

From Scheme 1 the rates are:

$$\frac{dN_{++}}{dt} = -(w_1^I + w_1^J + w_2)\,N_{++} + w_1^J\,N_{+-} + w_1^I\,N_{-+} + w_2\,N_{--}$$

$$\frac{dN_{+-}}{dt} = -(w_1^I + w_1^J + w_0)\,N_{+-} + w_1^J\,N_{++} + w_1^I\,N_{--} + w_0\,N_{-+}$$

$$\frac{dN_{-+}}{dt} = -(w_1^I + w_1^J + w_0)\,N_{-+} + w_1^J\,N_{--} + w_1^I\,N_{++} + w_0\,N_{+-}$$

$$\frac{dN_{--}}{dt} = -(w_1^I + w_1^J + w_2)\,N_{--} + w_1^J\,N_{-+} + w_1^I\,N_{+-} + w_2\,N_{++} \qquad \text{(A.1)}$$

The overall population differences ΔP^I and ΔP^J at any time are given by

$$\Delta P^I = (N_{++} + N_{+-}) - (N_{-+} + N_{--})$$

$$\Delta P^J = (N_{++} + N_{-+}) - (N_{+-} + N_{--}) \qquad \text{(A.2)}$$

from which the rates of change of ΔP^I and ΔP^J become

$$\frac{d\,\Delta P^I}{dt} = -(w_0 + 2w_1^I + w_2)\,\Delta P^I - (w_2 - w_0)\,\Delta P^J + \text{constant}^I$$

$$\frac{d\,\Delta P^J}{dt} = -(w_0 + 2w_1^J + w_2)\,\Delta P^J - (w_2 - w_0)\,\Delta P^I + \text{constant}^J \qquad \text{(A.3)}$$

The constants can be defined in terms of the equilibrium values $\Delta P^I(\infty)$ and $\Delta P^J(\infty)$ in such a way that $\frac{d\,\Delta P^I(\infty)}{dt} = \frac{d\,\Delta P^J(\infty)}{dt} = 0.$

$$\text{constant}^I = (w_0 + 2w_1^I + w_2)\,\Delta P^I(\infty) + (w_2 - w_0)\,\Delta P^J(\infty)$$

$$\text{constant}^J = (w_0 + 2w_1^J + w_2)\,\Delta P^J(\infty) + (w_2 - w_0)\,\Delta P^I(\infty) \qquad \text{(A.4)}$$

and therefore

$$\frac{d\,\Delta P^I}{dt} = (w_0 + 2w_1^I + w_2)\,(\Delta P^I(\infty) - \Delta P^I) + (w_2 - w_0)\,(\Delta P^J(\infty) - \Delta P^J)$$

$$\frac{d\,\Delta P^J}{dt} = (w_0 + 2w_1^J + w_2)\,(\Delta P^J(\infty) - \Delta P^J) + (w_2 - w_0)\,(\Delta P^I(\infty) - \Delta P^I) \qquad \text{(A.5)}$$

It is convenient to define $\rho_I = w_0 + 2w_1{}^I + w_2$, $\rho_J = w_0 + 2w_1{}^J + w_2$ and $\sigma_{IJ} = w_2 - w_0$, so that the final form of the equations is

$$\frac{d\,\Delta P^I}{dt} = \rho_I(\Delta P^I(\infty) - \Delta P^I) + \sigma_{IJ}(\Delta P^J(\infty) - \Delta P^J)$$

$$\frac{d\,\Delta P^J}{dt} = \rho_J(\Delta P^J(\infty) - \Delta P^J) + \sigma_{IJ}(\Delta P^I(\infty) - \Delta P^I) \qquad \text{(A.6)}$$

We define now $< I_z(t) >$ and $< J_z(t) >$ as the expectation value of I_z and J_z, respectively, at time t. Since ΔP^I is proportional to $< I_z(t) >$ (and ΔP^J to $< J_z(t) >$) the above equations can be rewritten as

$$\frac{d<I_z(t)>}{dt} = \rho_I(<I_z(\infty)> - <I_z(t)>) + \sigma_{IJ}(<J_z(\infty)> - <J_z(t)>)$$

$$\frac{d<J_z(t)>}{dt} = \rho_J(<J_z(\infty)> - <J_z(t)>) + \sigma_{IJ}(<I_z(\infty)> - <I_z(t)>) \qquad \text{(A.7)}$$

ρ is the intrinsic relaxation rate of the spin and σ is the cross-relaxation term.

References

[1] Bertini, I. and Luchinat, C. "NMR of Paramagnetic Molecules in Biological Systems" Benjamin/Cummings Menlo Park, CA, (1986).

[2] La Mar, G.N., Horrocks, W.D.Jr., Holm, R.H.Eds. "NMR of Paramagnetic Molecules" Academic Press New York, (1973).

[3] Bloembergen, N., Purcell, E.M., Pound, R.V. *Phys. Rev.* (1948) *73*, 679.

[4] Bloembergen, N. *J.Chem. Phys.* (1957) *27*, 575.

[5] Solomon, I. *Phys. Rev.* (1955) *99*, 559.

[6] Gueron, M. *J. Magn. Reson.* (1975) *19*, 58.

[7] Vega, A.J. and Fiat, D. *Mol. Phys.* (1976) *31*, 247.

[8] Noggle, J.H. and Schirmer, R.E. "The Nuclear Overhauser Effect" Academic Press New York, (1971).

[9] Neuhaus, D. and Williamson, M. "The Nuclear Overhauser Effect in Structural and Conformational Analysis" VCH Publ. New York, (1989).

[10] Overhauser, A.W. *Phys. Rev.* (1953) *89*, 689.

[11] Overhauser, A.W. *Phys. Rev.* (1953) *92*, 411.

[12] Woessner, D.E. *J. Chem. Phys.* (1965) *42*, 1855.

[13] Wagner, G. and Wütrich, K. *J. Magn. Reson.* (1979) *33*, 13, 675.

[14] Licoccia, S., Chatfield, M.J., La Mar, G.N., Smith, K.M., Mansfield, K.E. and Anderson, R.R. *J. Am. Chem. Soc.* (1989) *111*, 6087.

[15] Gordon, S.L. and Wütrich, K. *J. Am. Chem. Soc.* (1978) *100*, 7094.

[16] Wiliamson, M.P. and Neuhaus, D. *J. Magn. Reson.* (1987) *72*, 369.

[17] Johnson, R.D., Ramaprasad, S. and La Mar, G.N. *J. Am. Chem. Soc.* (1983) *105*, 7205.

[18] Ramaprasad, S., Johnson, R.D. and La Mar, G.N. *J. Am. Chem. Soc.* (1984) *106*, 5330.

[19] Mayer, A., Ogawa, S., Shulman, R.G., Yamane, T., Cavaleiro, J.A.S., Rocha-Gonsalves, A.M.d'A., Kenner, G.W. and Smith, K.M. *J. Mol. Biol.* (1974) *86*, 749.

[20] La Mar, G.N., Budd, D.L., Smith, K.M. and Langry, K.C. *J. Am. Chem. Soc.* (1980) *102*, 1822.

[21] Krishnamoorthi, R. Ph.D. Thesis, University of California, Davis (1982).

[22] Lecomte, T.J., Johnson, R.D. and La Mar, G.N. *Biochim. Biophys.Acta* (1985), 268.

[23] La Mar, G.N., Emerson, S.D., Lecomte, J.T.J., Pande, U., Smith, K.M., Craig, G.W. and Kehres, L.A. *J. Am. Chem. Soc.* (1986) *108*, 5568.

[24] Ramaprasad, S., Johnson, R.D. and La Mar, G.N. *J. Am. Chem. Soc.* (1984) *106*, 3632.

[25] Chatfield, M.J., La Mar, G.N., Parker, W.O., Smith, K.M., Leung, H.-K. and Morris, I.K. *J. Am. Chem. Soc.* (1988) *110*, 6352.

[26] Parker, W.O., Chatfield, M.J. and La Mar, G.N. *Biochemistry* (1989) *28*, 1517.

[27] Thanabal, V., De Ropp, J.S. and La Mar, G.N. *J. Am. Chem. Soc.* (1987) *109*, 265.

[28] Thanabal, V., De Ropp, J.S. and La Mar, G.N. *J. Am. Chem. Soc.* (1987) *109*, 7516.

[29] Satterlee, J.D., Erman, J.E. and De Ropp, J.S. *J. Biol. Chem.* (1987) *264*, 11578.

[30] Satterlee, J.D. and Moench, S.J. *Biophys. J.* (1977) *52*, 101.

[31] Satterlee, J.D., Moench, S.J. and Avizonis,D *.Biochim. Biophys. Acta* (1988) *952*, 317.

[32] Banci, L., Bertini, I. and Luchinat, C. *Structure and Bonding*, (1990) *72*, 113.

[33] Dugad, L.B., La Mar, G.N., Banci, L. and Bertini, I. *Biochemistry*, (1990) *29*, 2263.

[34] Bertini, I., Briganti, F., Luchinat, C. and Scozzafava, A. *Inorg. Chem.*, in press.

[35] Banci, L., Bertini, I., Luchinat, C. and Viezzoli, M.S. *Inorg. Chem.*, (1990) *29*, 1438.

[36] Banci, L., Bertini, I., Luchinat, C. and Piccioli, M. *Coord. Chem. Rev.*, (1990) *100*, 67.

[37] Bertini, I., Lanini, G., Luchinat, C., Messori, L., Monnanni, R. and Scozzafava, A. *J. Am. Chem. Soc.* (1985) *107*, 4391.

[38] Banci, L., Bertini, I., Luchinat, C., Piccioli, M., Turano, P. and Scozzafava, A. *Inorg. Chem.* (1989) *28*, 4650.

[39] Ernst, R.R., Bodenhausen, G. and Wokaun, A. "Principles of Nuclear Magnetic Resonance in One and Two Dimensions" Clarendon Press, Oxford, (1987).

[40] Croasmun, W.R. and Carlson, R.M.K. "Two dimensional NMR Spectroscopy" VCH Publ., New York, (1987).

[41] Jeener, J. *Ampere International Summer School*, Basko Poljie, Yugoslavia, (1971).

[42] Jeener, J., Meier, B.H., Bachmann, P. and Ernst, R.R. *J. Chem. Phys.* (1979) *71*, 4546.

[43] Macura, S. and Ernst, R.R. *Mol. Phys.* (1980) *41*, 95.

[44] Macura, S., Huang, Y., Suter, D. and Ernst, R.R. *J. Magn. Reson.* (1981) *43*, 259.

[45] Jenkins, B.G. and Lauffer, R.B. *Inorg. Chem.* (1988) *27*, 4730.

[46] Luchinat, C., Steuernagel, S., Turano, P. *Inorg. Chem.*, in press.

[47] Bothner-By, A.A., Stephens, R.L., Lee, J., Warren, C.D. and Jeanloz, R.W. *J. Am. Chem. Soc.* (1984) *106*, 811.

[48] Muller, L. and Ernst, R.R. *Mol. Phys.* (1979) *38*, 963.

[49] Braunschweiler, L. and Ernst, R.R. *J. Magn. Reson.* (1983) *53*, 521.

[50] Bax, A. and Davis, D.G. *J. Magn. Reson.* (1985) *63*, 207.

[51] Hartmann, S.R. and Hahn, E.L. *Phys. Rev.* (1962) *128*, 2042.
[52] Bax, A. and Davis, D.G. *J. Magn. Reson.* (1985) *65*, 355.
[53] Levitt, M.H., Freeman, R. and Freukiel, T. *J. Magn. Reson.* (1982) *47*, 328.

3. 3D NMR Spectroscopy in High Resolution NMR

Christian Griesinger

3.1 Introduction

This article intends to give a brief account about three-dimensional (3D) NMR spectroscopy and its applications in high resolution NMR spectroscopy to biomacromolecules. 3D NMR spectroscopy has been introduced only three years ago [1] and the results achieved so far are therefore only preliminary. However, trends have become obvious and we are now in a position to present the so far most successful and the most promising experiments as well as strategies for their interpretation together with examples. Two dimensional (2D) NMR spectroscopy has over the last decades revolutionized the field of NMR-spectroscopy [2]. The main reasons are the possibility to directly map structural information into a two dimensional spectrum. The structural information ranges from qualitative information such as connectivities of atoms in molecules, magnetic equivalence of sites and dynamic equilibrium to quantitative parameters like J-coupling constants and cross-relaxation rates which provide information about the local conformation such as distances and bond angles, as well as information about local mobility.

Yet, the success of two dimensional NMR spectroscopy relies also on the increased resolution inherent in the two dimensional display [3] that allowed for the first time to investigate large molecules in detail. In spectra of biopolymers with some hundred resonances originating from a restricted number of monomeric units that constitute the polymer the identification and assignment of resonances calls for 2D methods since overlap of cross-peaks in a 2D spectrum being characterized by two chemical shifts is clearly less probable than the overlap of signals in a 1D spectrum.

The basic 2D experiments that are applied to biomolecules are COSY [4], NOESY [5], TOCSY [6], ROESY [7], and hetero-COSY [8,9]. The principles of magnetization transfer they are based on are coherence transfer via scalar coupling (COSY and hetero-COSY), isotropic mixing (TOCSY), longitudinal cross-relaxation (NOESY) and transverse cross-relaxation (ROESY). They provide intraresidual (correlated spectra: COSY, TOCSY) as well as interresidual connectivity information (exchange experiments: NOESY, ROESY). A quantitative analysis of the given experiments yields coupling constants and cross-relaxation rates. An innumerable number of modifications has been derived over the last decade to adapt the specific NMR experiment to the given situation. Recent reviews may be consulted on this subject [10,11].

3.2 Limits of 2D spectroscopy

Limits of the two dimensional approach have become obvious occasionally. Consider the following hypothetical situation where overlap of resonances prohibits the measurement of certain interactions in the following spin system with five nuclei (Fig. 1).

Spin A is scalar coupled to B and C, spin D is dipolar coupled to E, B and C. Scalar coupling between two spins is denoted by a solid line, dipolar coupling by a broken line. In addition E and B have identical chemical shifts. The goal is to

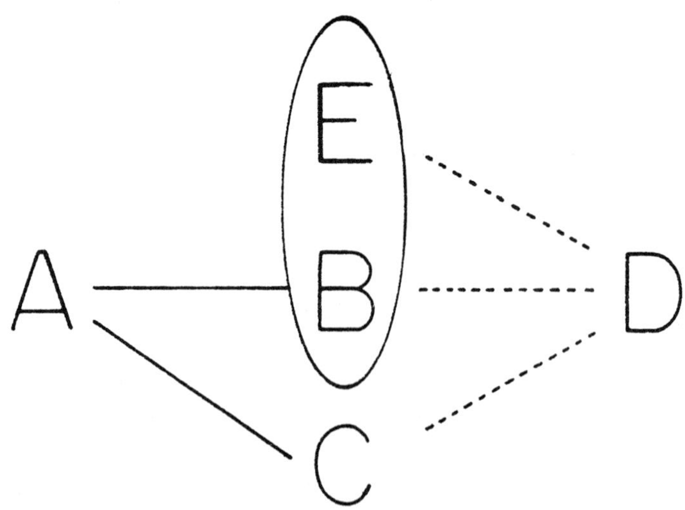

Fig. 1 Hypothetical five-spin system exhibiting scalar coupling of spin A to B and C (solid line) and dipolar coupling of D to B, C, and E (dotted line). The BD cross-relaxation rate is to be determined. B and E have the same chemical shifts.

measure the BD cross-relaxation rate which would normally be done e.g. in a 2D NOESY experiment, correlating the two chemical shifts of spins B and D. However, two frequencies as defined in the B, D-cross-peak in a NOESY experiment are not sufficient to distinguish this desired transfer from other possible transfers. Instead three frequencies have to be defined in the spectroscopic experiment: Ω_A, Ω_B, and Ω_D. Two frequencies are not sufficient: a cross-peak at $(\omega_1, \omega_2) = (\Omega_B, \Omega_D)$ could not be distinguished from a cross-peak at $(\omega_1, \omega_2) = (\Omega_E, \Omega_D)$. Such cross-peaks would result from a single transfer experiment (NOESY). Even a double transfer experiment such as a Relayed-NOESY [12] displaying cross-peaks at $(\omega_1, \omega_2) = (\Omega_A, \Omega_D)$ would not be sufficient because this cross-peak contains contributions from the A $\rightarrow$ B $\rightarrow$ D transfer as well as from the A $\rightarrow$ C $\rightarrow$ D transfer. Thus chemical shift information of three spins is required to measure the B/D cross-relaxation rate.

The definition of three frequency domains in a spectrum is necessary in this example only because of overlap of resonances. As molecules become larger, the probability of overlap of 2D cross-peaks increases leaving after the 2D analysis not only a few specific questions to solve. The introduction of a third frequency domain displaying chemical shifts then allows to spread the overlapping 2D peaks and thus identify the peaks and their chemical shifts unambiguously and derive the structural information. To demonstrate the limits of 2D spectroscopy the NOESY spectrum of calmodulin is shown as an example in Fig. 2. Even the identification of separate cross-peaks in this spectrum is impossible and only methods that increase the inherent resolution promise any progress.

The experimental realization of the frequency selection can be done either with selective pulses or with Fourier methods following the traits of 2D spectroscopy. Both methods have been proposed in the meantime in original papers: the Fourier approach of introducing all three frequency domains exclusively by the introduction of time domains [1,13-18], the exclusive use of selective pulses [19] and the mixed approach to use one selective pulse and two time domains [20]. The advantages and disadvantages of the definition of frequency domains by the Fourier principle or by the use of selective pulses will be discussed later.

The use of three frequencies in NMR experiments is not completely new. Experiments in NMR with three frequency domains in NMR have been performed occasionally in triple resonance studies [21]. 2D experiments using weak CW irradiation lead also to three frequency variables [22]. Several 2D experiments contain already a third time parameter which could be converted into a time variable for 3D spectroscopy. This applies in particular to experiments that involve two coherence transfer steps, separated by a time delay, such as relay experiments [12,23], or an extended cross-relaxation or exchange period, such as NOESY and EXSY. Indirectly detected 2D spectra, such as 2D correlation in zero field with high field detection [24,25] require a 3D experiment. Leaving the field of analytical NMR, 3D and even higher dimensional NMR experiments are performed in the connection with imaging. Finally, we should note, that also chemically new frequency domains can be introduced. Site selective or residue selective isotope labelling that becomes more and more popular with the general availability of genetic engineering techniques produce 2D spectra that are considerably simplified based on the "selective excitation" of certain residues or residue types in the molecule [e.g. 26].

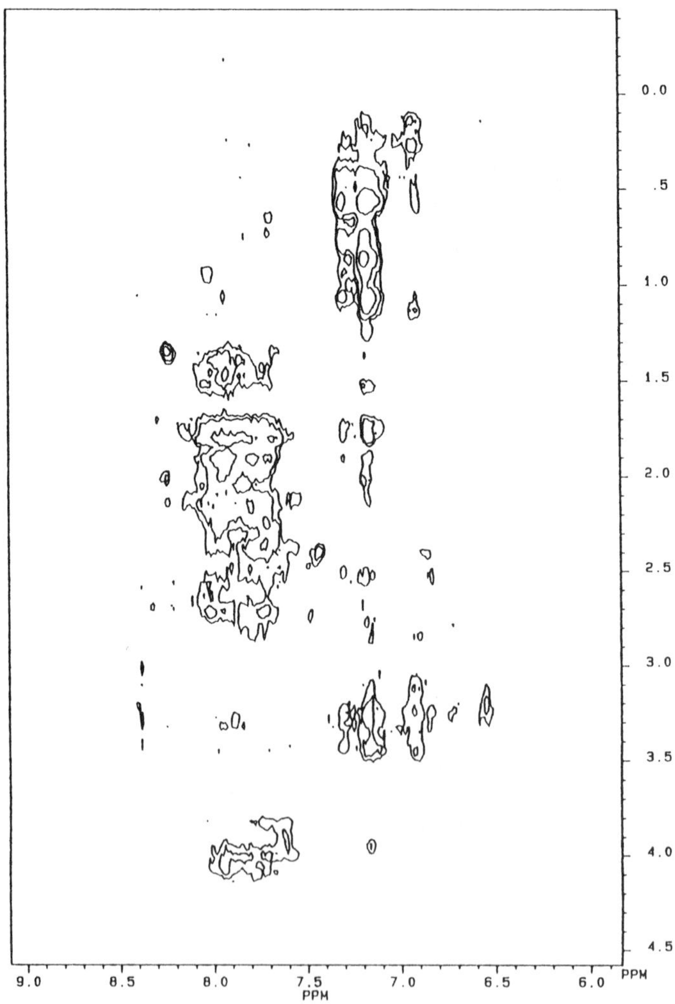

Fig. 2 H_α, NH region of the NOESY spectrum of calmodulin in H_2O / D_2O 9/1 at pH = 5. It demonstrates the severe overlap of resonances making it impossible even to identify separate cross-peaks.

3.3 Construction scheme of 3D experiments

3D time domain NMR experiments are constructed in close analogy to 2D pulse experiments (Fig. 3). An additional evolution period-mixing period element is inserted after the preparation pulse leaving one preparation, two evolution, two mixing and one detection periods. The preparation and mixing sequences are constructed from pulses and delays. We may also consider a 3D pulse sequence as the combination of two 2D experiments by omission of the detection period of the first and the preparation period of the second experiment. Chemical shifts are most appropriate to disperse cross-peaks in a spectrum. Thus correlation of three chemical shifts in the three frequency domains by coherence transfer between the three involved spins will lead to the most useful experiments which we henceforth call transfer experiments. These experiments are based on COSY, NOESY, TOCSY, ROESY, and hetero-COSY being the fundamental 2D transfer experiments. Useful 3D pulse sequences are compiled in Fig. 4. They employ non-selective as well as selective pulses [27] for volume selection in the 3D frequency space as explained in the course of the text. The selective pulses used in these experiments excite whole regions in a spectrum which is to be contrasted with selective pulses which excite only one resonance. Details of the pulse sequences are found in the caption of Fig. 4.

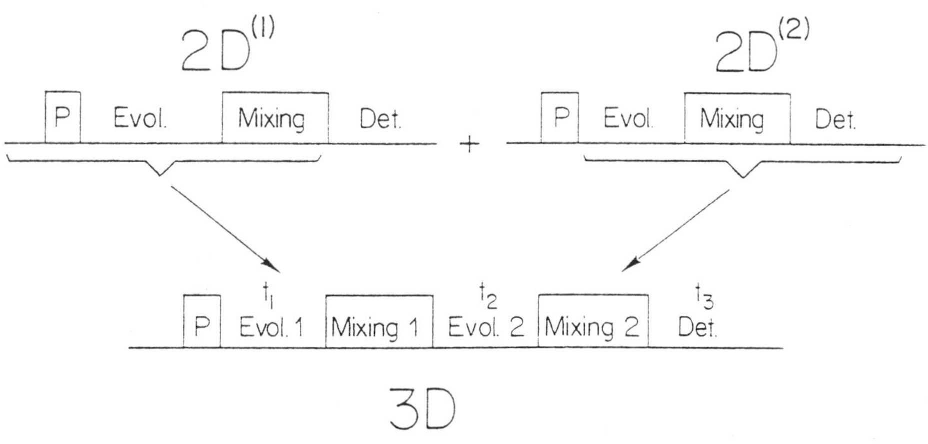

Fig. 3 Experimental implementation of defining three frequency axes: the pure Fourier approach introduces the three frequency domains via independently incremented time domains in complete analogy to 2D spectroscopy. The frequencies of the coherences during t_1 and t_2 are obtained by Fourier transformation along t_1 and t_2. The two mixing processes may be different. The 3D experiment is obtained by merging two 2D sequences.

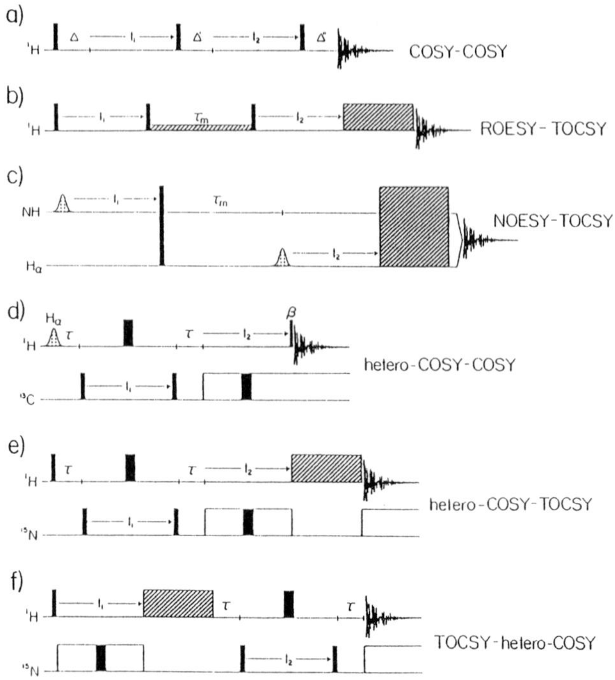

Fig. 4 Useful pulse sequences in 3D time domain NMR spectroscopy. Non selective pulses are indicated by filled bars. $\pi/2$ and π pulses are distinguished by their widths, pulses with flip angles deviating from $\pi/2$ or π are specially indicated. Frequency selective pulses, all nominally $\pi/2$, are drawn with dotted Gaussian shape. They cover usually a whole range of resonances, e.g. the NH region. They must not be confused with multiplet selective pulses. a) COSY-COSY sequence with delays in the evolution times to allow for partial refocusing of the magnetization. b) Non-selective ROESY-TOCSY pulse sequence. Both ROESY and TOCSY mixing sequences are represented by hatched areas, the ROESY mixing by a lower one due to the lower power used in this sequence. c) NOESY-TOCSY pulse sequence with selective excitation in ω_1 and ω_2. Because the non selective pulse after t_1 excites all resonances, phase cycling is required to select zero quantum coherence during the mixing time τ_m. The asterisk indicates a refocusing sequence consisting of a non-selective π pulse followed by a delay Δ of half the length of the selective pulse (see text). d) Heteronuclear COSY-COSY sequence exemplified by ^{1}H, ^{13}C correlation. In applications to peptides and proteins, the selective $\pi/2$ pulse may, for example, be adjusted to excite the H_α resonance region. No further selective pulses are required for frequency restriction in ω_2. The delay τ must be tuned to $(2^1J_{CH})^{-1}$ for optimum sensitivity. The first τ delay includes half the duration of the selective pulse. If possible, ^{13}C decoupling in the C_α region is applied during t_1 and during acquisition. Alternatively, a central π pulse can ensure decoupling at least during t_2. e) Heteronuclear COSY-TOCSY sequence arranged for ^{1}H, ^{15}N correlation. In peptides and proteins, the frequency range in t_2 is restricted to amide protons without any selective pulses. The selectivity is achieved by tuning the τ delays to the large one-bond couplings ($\tau = (2^1J_{NH})^{-1}$). Note that no ^{15}N decoupling is applied during the homonuclear TOCSY mixing in order to avoid a sensitivity loss caused by heteronuclear magnetization transfer. f) TOCSY-heteronuclear COSY sequence arranged for ^{1}H, ^{15}N correlation. In peptides and proteins, the order of the two constituent 2D sequences allows the observation of the NH resonances in ω_3, which solves the problem of water suppression on the expense of the resolution in ω_2 because the total proton spectral width has to be covered in the experiment.

3.4 Classification of peaks in 3D spectra

In 2D transfer experiments which transfer coherence between two spins during the mixing period we distinguish cross-peaks and diagonal peaks originating from the transfers:

B → A and B → B, respectively.

A and B denote two different spins. These are the only two peak types. While diagonal peaks in the 2D spectrum have the same resolution as a 1D spectrum and usually are not useful for the interpretation, cross-peaks carry the new information and have a much increased resolution. 2D spectra originating from the most important 2D experiments exhibit in addition symmetry about the $\omega_1 = \omega_2$ axis [28]. In other words, they have the property: $M_{AB} = M_{BA}$, where M_{AB} is the transfer efficiency of the AB transfer.
In 3D experiments there are four types of peaks:

Diagonal Peaks	B	→	B	→	B	with $\omega_1=\omega_2=\omega_3$
Cross Diagonal Peaks	B	→	B	→	A	with $\omega_1=\omega_2$
Cross Diagonal Peaks	B	→	A	→	A	with $\omega_2=\omega_3$
Back Transfer Peaks	B	→	A	→	B	with $\omega_1=\omega_3$
and Cross-Peaks	B	→	A	→	C	

Only the latter two types of peaks bear information that is not contained in 2D experiments. The three planes $\omega_1 = \omega_2$, $\omega_2 = \omega_3$ and $\omega_1 = \omega_3$ are called, according to the peaks which lie on them, cross diagonal planes and back-transfer planes (Fig. 5). The resolution in the cross-diagonal and back-transfer planes is the same as in 2D spectra, because only two chemical shift define the location of these peaks. This has consequences for the strategies of interpreting 3D spectra.

3D spectra lack global symmetry analogous to the reflection symmetry at $\omega_1 = \omega_2$ in 2D spectra except for the combination of two 2D sequences that individually lead to $\omega_1 = \omega_2$ reflection symmetric 2D spectra. Such 3D spectra are reflection symmetric about the back-transfer plane $\omega_1 = \omega_3$. An example is e.g. a COSY-COSY experiment (Fig. 4a) or a EXSY-EXSY experiment. In the following we will learn that the combination of different types of 2D experiments forms the most interesting 3D experiments. They lack any global symmetry. The redundancy of 2D due to symmetry is lost in 3D.

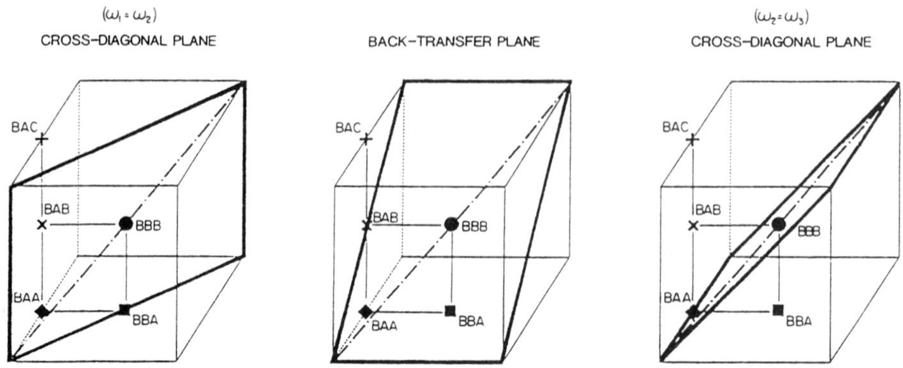

Fig. 5 Peak types in 3D NMR spectra for three resonances A, B, and C. The diagonal peak BBB, the cross diagonal peaks BBA and BAA, the back-transfer peak BAB and the cross-peak BAC are given in each of the three cubes. In addition, one of the three diagonal planes is indicated. The peaks which have identical shifts at least in two dimensions consequently lie on the corresponding plane.

3.5 Information Content of 3D spectra

3D transfer spectra contain as a consequence of the 3D sequence construction scheme only coherence transfers that are also found in the corresponding two 2D experiments, the 3D experiment is constructed from. There are no new physical effects that could be observed in 3D spectroscopy. This is true even for 3D experiments that correlate e.g. double quantum coherences in ω_1 and ω_2 which is impossible to do in a conventional 2D experiment. The advantage of the 3D spectrum that remains is the additional frequency coordinate which is valuable to increase resolution and separate various contributions to cross-peaks arising from different coherence transfer paths. To illustrate these claims consider a 3D experiment that is constructed from two 2D experiments $2D^{(1)}$ and $2D^{(2)}$. The coherence | rs> is transferred to | tu> in the first mixing sequence and the coherence | tu> is transferred to | vw> in the second experiment:

$$2D^{(1)}: \quad |rs\rangle \quad \rightarrow \quad M_1\,|tu\rangle$$

$$2D^{(2)}: \quad |tu\rangle \quad \rightarrow \quad M_2\,|vw\rangle$$

Omission of the second evolution time in the 3D experiment would result in the same transfer showing that a 2D experiment with a mixing period constructed from the two mixing periods of the 3D experiment exploits the same transfer pathways. However, the frequency information about the intermediate coherence is lost. If there exist several pathways from | rs> to | vw> they cannot be disentangled in the 2D spectrum in contrast to the 3D experiment. The three frequency coordinates guarantee inherently higher resolution as pointed out earlier. Thus qualitative information which manifests itself in the absence or presence of a cross-peak is easier to extract from 3D spectra than from 2D spectra, because the identification of peaks is easier.

One important source of quantitative information in 2D experiments are cross-peak integrals, the extraction of cross-relaxation rates in 2D NOESY or ROESY is based on. If we denote the respective transfer coefficients of the mixing processes in the first and second 2D experiment by M_1 and M_2 the cross-peak at $\omega_1 = \Omega_{rs}$, $\omega_2 = \Omega_{tu}$ and $\omega_3 = \Omega_{vw}$ in the corresponding 3D experiment has consequently an intensity $M_1 * M_2$. Ω_{rs} denotes the eigenfrequency of the coherence | rs>. Since only products of transfer coefficients can be measured, quantitative information is less easy to extract. The separation of the two transfer coefficients is not trivial. However, if one of the coefficients is uniform and spin system independent, e.g. M_1 = const, the transfer efficiency depends exclusively on M_2, which represents quantitative 2D information. This situation occurs in 3D experiments which contain a heteronuclear transfer step via heteronuclear ^1J couplings (Fig. 4d-f). By contrast, homonuclear transfer experiments like COSY, TOCSY, ROESY, and NOESY introduce transfer amplitudes that strongly depend on the spin system. They are difficult to measure or estimate from previous knowledge. Thus, heteronuclear experiments are especially suited to obtain quantitative information from 3D spectra.

3.6 Sensitivity of 3D spectra

For practical considerations the sensitivity of 3D spectroscopy in comparison to 2D experiments is important to know in order to estimate the feasibility under given circumstances of instrument time, sample concentration and sensitivity of the instrument. The sensitivity of the most interesting experiments which transfer exclusively in-phase magnetization is easily calculated if we assume unresolved multiplets of Lorentzian shape in all dimensions of a width $(\pi T_2)^{-1}$. This situation is common for macromolecules. We then obtain for the time domain signal for the ABC cross-peak:

$$S_{3D}(t_1,t_2,t_3) = M_{AB}\, M_{BC}\, e^{-t_1/T_2^A}\, e^{-t_2/T_2^B}\, e^{-t_3/T_2^C}$$

$(\pi T^A{}_2)^{-1}$ is the multiplet width of resonance A. M_{AB} and M_{BC} are the transfer coefficients. For the signal in the two constituent 2D experiments 2D $^{(1)}$ and 2D $^{(2)}$ we find:

$$S_{2D}{}^{(1)}(t_1,t_2) = M_{AB}\, e^{-t_1^{(1)}/T_2^A}\, e^{-t_2^{(1)}/T_2^B}$$

$$S_{2D}{}^{(2)}(t_1,t_2) = M_{BC}\, e^{-t_1^{(2)}/T_2^B}\, e^{-t_2^{(2)}/T_2^C}$$

For a 1D spectrum with a $\pi/2$ excitation pulse and with observation of the B spin one obtains the signal:

$$S_{1D}(t) = e^{-t/T_2^B}$$

Taking these four formulae together one obtains the ratio of the product of the signal in the 3D spectrum and the signal in the 1D spectrum compared to the product of the signal in the two constituent 2D spectra:

$$\frac{S_{3D}\, S_{1D}}{S_{2D}{}^{(1)}\, S_{2D}{}^{(2)}} = 1 \qquad \text{for } t_1 = t_1^{(1)} \text{ and } t_2 = t_1^{(2)}$$

Since the noise amplitude during the detection in the four experiments is the same this result holds also for the comparison of the Signal to Noise ratios per square root of unit time. A more rigorous treatment that allows also for different maximum evolution times in the 3D sequence and the two 2D experiments gives for matched filtering [2]:

$$\frac{S/N_{3D}^{ABC}\;\; S/N_{1D}^{B}}{S/N_{2D}^{AB(1)}\;\; S/N_{2D}^{BC(2)}} = \frac{(1-e^{-2t_1^{max}/T_2^A})\;(1-e^{-2t_2^{max}/T_2^B})\;t_1^{(1)max}\;\;t_1^{(2)max}}{(1-e^{-2t_1^{(1)max}/T_2^A})\;(1-e^{-2t_2^{(2)max}/T_2^B})\;t_1^{max}\;\;t_2^{max}}$$

The factor on the right is normally > 1 since the evolution times in 3D experiment, t_1 and t_2, are usually smaller than the evolution times in 2D experiments, $t_1{}^{(1)max}$ and $t_1{}^{(2)max}$. The signal to noise ratios are in units of square root of inverse measurement time.

We give an example:

Consider the example of a ^{15}N enriched protein. The selected spectrum is a ^{15}N, H-hetero-COSY-TOCSY experiment. The ^{15}N, H-correlation takes 1 h to achieve a signal to noise ratio of 10. So the signal to noise ratio per square root unit time is 10 $h^{-1/2}$. The signal to noise for a 1D spectrum is 5 after 10 min and the S/N of the TOCSY is 5 $(16\ h)^{-1/2}$. So we arrive at a signal to noise per square root unit time:

$$S/N_{3D} = \frac{5\,(16h)^{-\frac{1}{2}}\,10\,h^{-\frac{1}{2}}}{5\,(10min)^{-\frac{1}{2}}} = 10\,(96h)^{-\frac{1}{2}} = 5\,(24h)^{-\frac{1}{2}}$$

So the signal to noise of the 3D experiment will be 10 after 96 hours or 5 after 24 hours.

3.7 Practical Aspect of 3D Spectroscopy

The experimental realization of the frequency domains in NMR can be done with selective pulses which excite only one resonance or the Fourier principle by the introduction of an evolution time. Selective pulses have the advantage that the resolution that can be obtained is independent of the experiment time whereas the resolution that can be achieved for a given spectral width with the Fourier principle depends linearly on the experiment time. The disadvantage of selective pulses is their restricted selectivity because relaxation reduces the intensity of the signals the more selective the pulse is. Even severer is the lack of the multiplex advantage. Thus selective pulses will be employed in cases where only few resonances have to be selected and the signal to noise ratio of the sample is high. Fourier spectroscopy however is advantageous for low sensitivity samples with many resonances. The latter case clearly applies to macromolecules. The reasons to choose 3D time domain spectroscopy are the same reasons that forced people to substitute series of double resonance experiments with COSY. Consequently, all 3D NMR experiments on biomolecules have been performed with Fourier methods. The two evolution time domains in 3D time domain spectroscopy have to be incremented independently. This makes 3D experiments time consuming. Good digitization like in 2D spectra cannot be achieved in a reasonable amount of time as the following example illustrates: 5 Hz digital resolution for a normal

3.7.2 b) Selective Approach

In order to optimize resolution or reduce measurement time, frequency selection with selective excitation and mixing sequences is employed. Frequency selective excitation is done by comparably long shaped pulses (usually Gaussian or Hermitian [27,29]) whose widths are adjusted to the frequency range they are supposed to excite. The realization of selective mixing depends on the specific mixing scheme that is applied. Selective mixing in COSY employs two preferably simultaneous pulses on two non-overlapping spectral regions [1]. For the selective NOESY experiment, the first mixing pulse can be nonselective and the second is selective (Fig. 4c). Phase gradients that lead to distorted lineshapes introduced by the selective pulses can be removed by refocusing with a π–Δ sequence following the selective pulse if it transforms longitudinal magnetization into transverse. If the selective pulse achieves the opposite transfer a Δ–π sequence should precede the selective pulse. Δ is approximately half the duration of the selective pulse. The application of self refocused selective pulses is an alternative solution [30].

The number of selective pulses should be kept as small as possible because of intensity losses during the selective pulses and the necessity to introduce refocusing pulses that have to be phase cycled when in phase magnetization prevails during the respective time domain.

As an example the NOESY-TOCSY spectrum of buserilin is given . The pulse sequence in Fig. 4c has been used. The NH resonances are selectively excited. Selective NOESY-mixing to the H_α-resonances follows and finally TOCSY-mixing to all protons in the spin system (Fig. 7). NOESY-TOCSY has been applied successfully also to the 45-aminoacid containing protein α-purothionein [14].

Heteronuclear experiments may be implemented with or without selective pulses. Correlation with carbons requires frequency selection because of the huge chemical shift range. For natural abundance compounds this has to be done with selective pulses. Samples with isotope enrichment at certain positions with restricted chemical shift range (e.g in the C_α region in a peptide) exhibit a built-in frequency selection and can be treated with non-selective pulses. ^{15}N in peptides and proteins is in this respect well behaved, since it has a restricted chemical shift range. Combinations of the hetero-COSY sequence [8,9] with homonuclear transfer experiments can be used to optimize intensity of the heteronuclear transfer step (Fig. 4e) [18]. The order of the two constituent experiments should be reversed in ^{15}N-correlations if the H_α-resonances should be observed while they would be buried under the water resonance in ω_3. Then first the homonuclear

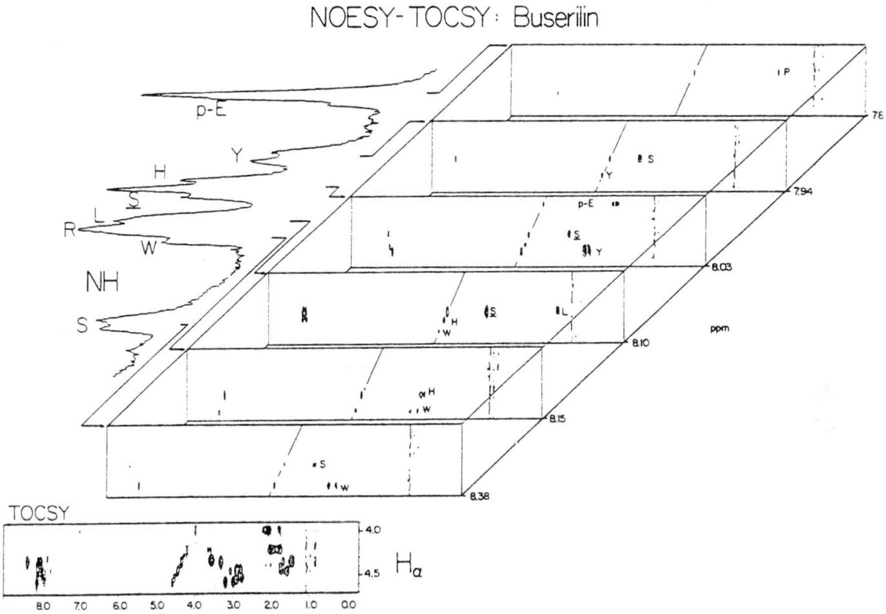

Fig. 7 300 MHz selective 3D NOESY-TOCSY experiment of buserilin in DMSO-d_6 with the selection of the NH, H_α, complete spectrum part. The data matrix of 96*96*2K was zero-filled to 256*256*4K prior to Fourier transformation. 32 scans were performed per t_1, t_2 value. The NOESY mixing time was 250 ms, the TOCSY mixing, achieved with MLEV-17 [35,36] lasted 42.66 ms. The duration of one scan was 2.66 sec. Selective excitation at the beginning of t_1 and t_2 was achieved with a 10.0 ms Gaussian pulse followed by a non-selective π pulse and a 4.0 ms delay. (ω_2, ω_3) slices are taken at interesting positions in ω_1 which are marked by arrows in the 1D spectrum of the NH resonances. The enhanced resolution in the TOCSY sections of the 3D spectrum is obvious by comparison with the 2D TOCSY spectrum of buserilin at the bottom of the figure. The assignment of the peaks is given with the one-letter codes for the amino acids. *Reprinted from Ref. 15.*

transfer step to the NH resonances is performed and then the hetero-COSY transfer (Fig. 4f). Decoupling during t_3 increases the sensitivity by a factor of 2.

As an example for heteronuclear experiments the ^{13}C, H-hetero-COSY-COSY (Fig. 4d) spectrum of buserilin is shown in Fig. 8. $^{13}C_\alpha$, H_α and all spins coupling to the H_α's are correlated in the experiment. A stereographic view shows the $^{13}C_\alpha$, H_α, NH part of the spectrum together with the assignment of the peaks.

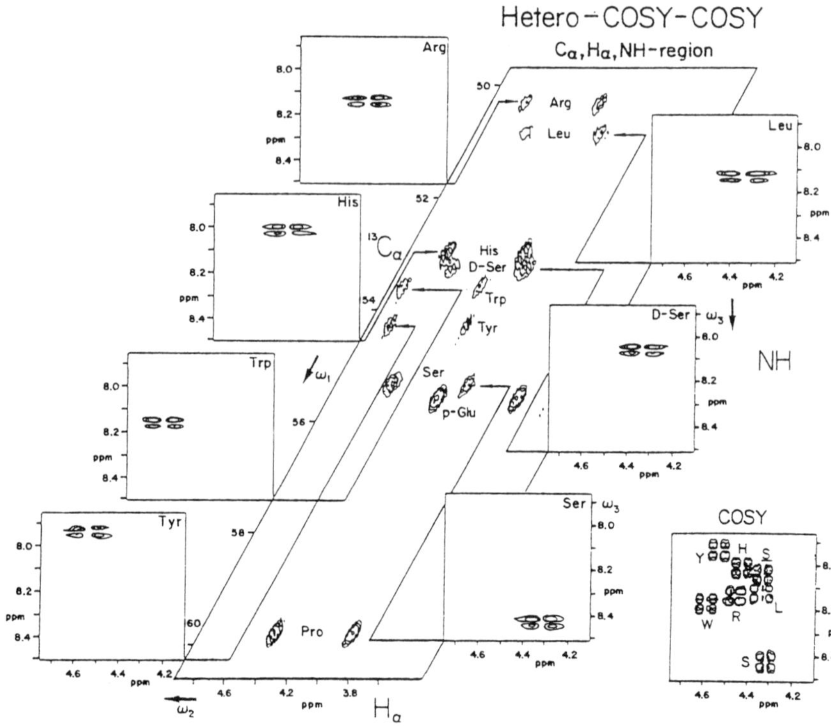

Fig. 8 300 MHz 3D hetero-COSY-COSY of buserilin acquired with pulse sequence of Fig. 4f correlating $^{13}C_\alpha$, H_α, and NH resonances in ω_1, ω_2, and ω_3, respectively. Prior to the actual pulse sequence, selective inversion (BIRD-sequence [37]) of protons not bound to ^{13}C, followed by a 320 ms delay, served to zero the magnetization of the NH resonances. 96*57 t_1, t_2 experiments, each with 32 scans and 2K data acquisition in t_2 were performed. The delay τ was tuned to $^1J_{CH}$ = 143 Hz. The duration of one scan was 2.5 sec. Selective excitation of the H_α resonances was achieved with a 6.5 ms Gaussian pulse. The data matrix was zero-filled to 256*128*4K points prior to Fourier transformation. The interesting NH region in ω_3 was cut out in order to accommodate the data on disk. Assignments are given with the one letter codes for amino acids. *Reprinted from Ref. 15.*

3.8 Extraction of Information

The information that is needed to answer interesting biochemical questions with NMR is the assignment of resonances. This is a prerequisite no matter whether the structure of the compound shall be derived from the analysis of quantitative information or whether related compounds shall be compared or binding of substrates to macromolecules shall be investigated. Since qualitative information is easily extracted from 3D spectra, we will design in the following assignment strategies which use the high resolution in 3D optimally.

3.8.1 Sequential assignment

Since all biomolecules are macromolecules built from a small number of different monomers the same monomer occurs several times in the sequence of the macromolecule. Therefore sequential assignment of residues is indispensable for the complete assignment of the resonances. This requires usually intraresidual as well as interresidual connectivity information. Sequential analysis in 2D NMR of peptides and proteins from combinations of NOESY (ROESY) and COSY (TOCSY) is done via sequential assignment of proton resonances along the whole backbone chain. The shortest sequential distances are found between NH_{i+1} and $H_{\alpha,i}$ for β-sheet structures (2.2 Å) and NH_i and NH_{i+1} for helical structures (2.7 Å). Consequently, sequential assignment procedures are based on both the intraresidual $NH_i / H_{\alpha,i}$ connectivities and the sequential $H_{\alpha,i} / NH_{i+1}$ NOE's or directly the NH_i / NH_{i+1} NOE's. Simultaneous evaluation of incoherent and coherent experiments can be demonstrated schematically on the COSY/NOESY plot for β-sheet structures [31] in Fig. 9.

As the assignment procedure relies on lines which hit cross-peaks and thus identify the sequential connectivity the 2D assignment procedure hinges on the non-degeneracy of the NH and H_α resonances. Resonances being degenerate at the resolution of the 2D spectrum make the assignment ambiguous.

3D assignment procedures are preferably based on lines in 3D space. Starting from a cross-peak one preferably finds the next cross-peak by keeping fixed *two* of the chemical shifts and varying the third frequency thus searching along a *line* in 3D space. Then the assignment is unambiguous as long as there are no degenerate *pairs* of chemical shifts in the 2D spectrum. This is less likely than overlap of resonances in the 1D spectrum.

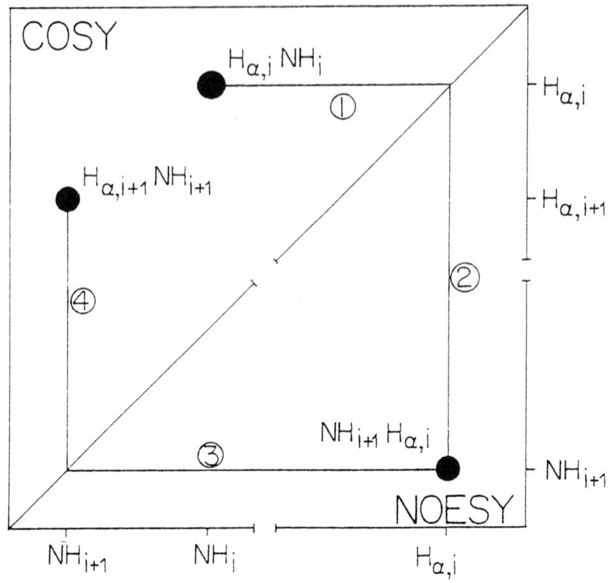

Fig. 9 Schematic sequence analysis procedure in 2D based on simultaneous evaluation of COSY and NOESY. The cross-peaks are connected by lines thus keeping one frequency coordinate fixed. Starting with the COSY cross-peak $H_{\alpha,i}$, NH_i a line along ω_1 is defined through $\omega_2 = \Omega_{NHi}$. Reflection of this line at $\omega_1 = \omega_2$ yields a new ω_2 line at $\omega_1 = \Omega_{NHi}$ which hits the NOESY cross-peak NH_i, $H_{\alpha,i-1}$. This in turn defines a new ω_1 line at $\omega_2 = \Omega_{H\alpha,i-1}$ which upon reflection at $\omega_1 = \omega_2$ hits the COSY cross-peak $H_{\alpha,i-1}$ NH_{i-1}. This last step terminates one sequential assignment cycle.

Based on NH_i / NH_{i+1}-distances for helices and NH_i/$H_{\alpha,i+1}$-distances for β-sheet structures we can formulate the following assignment procedures (Table 1).

Tab. 1 Assignment procedures based on nonselective 3D experiments*.

	$NH^{(i)}$–$H_\alpha^{(i)}$	...	$NH^{(i+1)}$–$H_\alpha^{(i+1)}$		...	$NH^{(i+2)}$–$H_\alpha^{(i+2)}$	
	1		2	3			
β-sheet			3	2		1	
				1		2	3
	1		2	3			
helix			2	3		1	
			1			2	3

	$NH^{(i)}$–$^{15}N^{(i)}$	...	$NH^{(i+1)}$–$^{15}N^{(i+1)}$		...	$NH^{(i+2)}$–$^{15}N^{(i+2)}$	
	1		2	3			
helix			2	3		1	
			1			2	3

* The number 1, 2, 3 indicate the frequency coordinates ω_1, ω_2, ω_3 of 3D cross-peak.

H_α, NH and NH, ^{15}N, respectively, constitute the chain of resonances. The numbers in the table refer to the frequency coordinate where a certain resonance appears. A 3D peak is defined by three numbers in a row. From the table we can learn that for the β-sheet the H_α / NH / H_α and NH / H_α / NH regions are required in a NOESY-TOCSY experiment. For the helical parts the NH / NH / H_α or the NH / NH / ^{15}N are necessary in a NOESY-TOCSY or hetero-COSY-NOESY experiment, respectively. These spectral regions can be recorded with selective experiments or with one nonselective experiment.

All three assignment procedures use one dimensional search to proceed from one cross-peak to the next one. As an example the β-sheet assignment procedure shall be exemplified with the nonselective ROESY-TOCSY spectrum of Fig. 6 for the tripeptide subsequence Ser-Tyr-D-Ser in buserilin (Fig. 10 only the interesting H_α / NH / H_α and NH / H_α / NH regions are shown). The four relevant peaks are summarized in Table 2.

Tab. 2

	S			Y			<u>S</u>		
	$NH^{(i)}$ –	$H_\alpha{}^{(i)}$	...	$NH^{(i+1)}$ –	$H_\alpha{}^{(i+1)}$	...	$NH^{(i+2)}$ –	$H_\alpha{}^{(i+2)}$	...
Line 1	3	2		1					
Line 2		1		2	3				
Line 3				3	2		1		
Line 4					1		2	3	

We start from the Y_{NH} / S_α / S_{NH} cross- peak. It defines the white line [1] along ω_3 at $\omega_1 = \Omega_{Y(NH)}$, $\omega_2 = \Omega_{S(\alpha)}$. Reflection of this line at the $\omega_1 = \omega_2$ diagonal plane produces the white line [2] at $\omega_1 = \Omega_{S(\alpha)}$, $\omega_2 = \Omega_{Y(NH)}$, which hits the cross-peak S_α / Y_{NH} / Y_α. This cross-peak defines the red line [3] now along ω_1 at $\omega_2 = \Omega_{Y(NH)}$, $\omega_3 = \Omega_{Y(\alpha)}$. Upon reflection at $\omega_2 = \omega_3$ the red line [4] at $\omega_2 = \Omega_{Y(\alpha)}$, $\omega_3 = \Omega_{Y(NH)}$ is defined which hits the $\underline{S}_{NH}$ / Y_α / Y_{NH} cross-peak. Thus one sequencing step is finished and the successive application of this procedure allows a complete sequence analysis. Fig. 10 shows the application of the sequencing procedure at the H_α / NH / H_α and NH / H_α / NH region of the ROESY-TOCSY spectrum of buserilin in a stereographic view. The four lines constituting the Ser / Tyr and Tyr / D-Ser connectivity are indicated in white and red respectively.

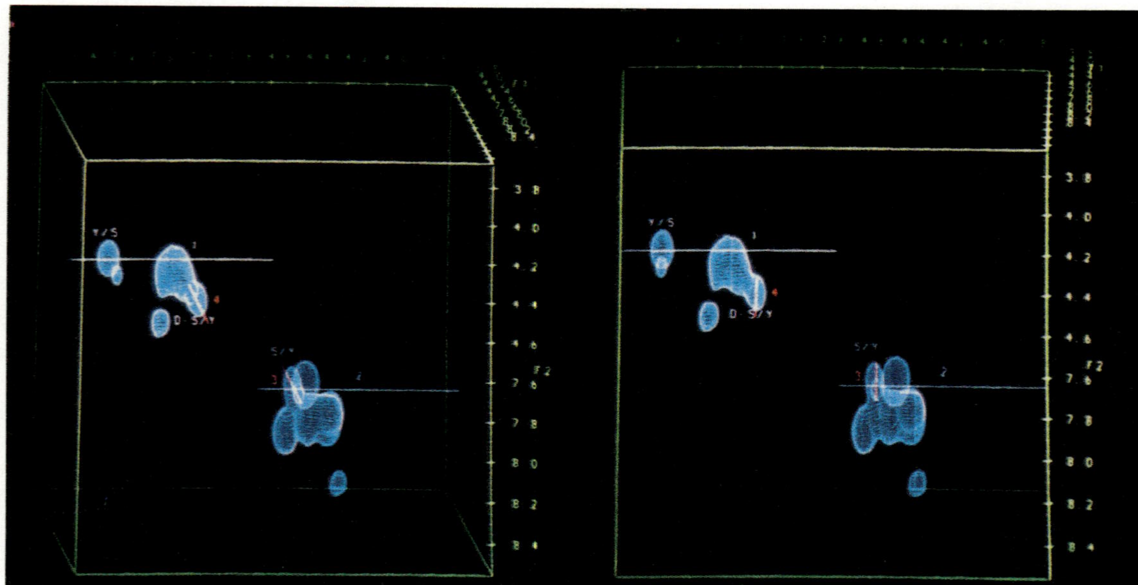

Fig. 10 3D sequence analysis in the NH, H_{α}, NH and H_{α}, NH, H_{α} region of the ROESY-TOCSY spectrum of buserilin taken from the spectrum described in Fig. 6. Each of the two regions has a volume of 1 ppm^3 and is digitized with 64*64*128 points in ω_1, ω_2 and ω_3 respectively. The three cross- peaks that are correlated by the assignment procedure are indicated in white with the one-letter code for amino acids. Four lines illustrate the path of the assignment procedure as described in detail in the text.

Assignment strategies that cannot use the ideal 3D sequence analysis procedure are discussed in Table 3.

Tab. 3

	$^{15}N^{(i-1)}$–	$NH^{(i-1)}$–	$H_\alpha{}^{(i-1)}$ ...	$^{15}N^{(i)}$–	$NH^{(i)}$–	$H_\alpha{}^{(i)}$ ...	$^{15}N^{(i+1)}$–	$NH^{(i+1)}$–	$H_\alpha{}^{(i+1)}$
NH,Hα,NH region of NOESY-TOCSY		3	2		1				
					1,3	2			
					3	2		1	
Hα,NH,Hα, region of NOESY-TOCSY			1		2	3			
					2	1,3			
						1		2	3
a	1	2	3						
b			3	1	2				
a				1	2	3			
b						3	1	2	

a Intraresidual cross-peak in ^{15}N, H COSY-TOCSY or ^{15}N, H-COSY-NOESY.

b Intraresidual cross-peak in ^{15}N, H-COSY-NOESY.

Here, sequence analysis procedures are presented for β-sheet domains when only one region of the NOESY-TOCSY spectrum is available. Possible reasons are the application of only one selective experiment or the water resonance burying H_α-resonances in ω_3. In this case back-transfer peaks have to be employed to obtain a formally equivalent assignment procedure as in the above discussed case where only one frequency coordinate is changed from one cross-peak to the next. A schematic representation is given in Fig. 11. However, the resolution in the back-transfer plane is practically the same as in a conventional 2D experiment recorded with the same resolution as the 3D experiment making this procedure less robust against degeneracies of resonances.

An example is given in the NOESY-TOCSY experiment of buserilin at the Trp-Ser-Tyr (W-S-Y) fragment (Fig. 12). The $W_{NH} / W_\alpha / W_{NH}$ back transfer peak and the $S_{NH} / W_\alpha / W_{NH}$ cross-peak which lie in the same ω_1, ω_2-plane have the same ω_2, ω_3-coordinates and are easily identified to belong together. Following now a line along ω_3 to the back-transfer plane and from there a line along ω_2 the $S_{NH} / S_\alpha / S_{NH}$ back-transfer peak is found. From there the $Y_{NH} / S_\alpha / S_{NH}$ cross-peak is hit following a line along ω_1 keeping fixed the ω_2 and ω_3 coordinates. Note that the other peaks in this plane do not disturb the identification of the $Y_{NH} / S_\alpha / S_{NH}$ cross-peak. With the connectivities of the two cross-peaks and the two back-transfer peaks the W-S-Y tripeptide fragment has been established.

For heteronuclear experiments based on hetero-COSY-NOESY and hetero-COSY-TOCSY only one of the assignment steps namely the correlation of the

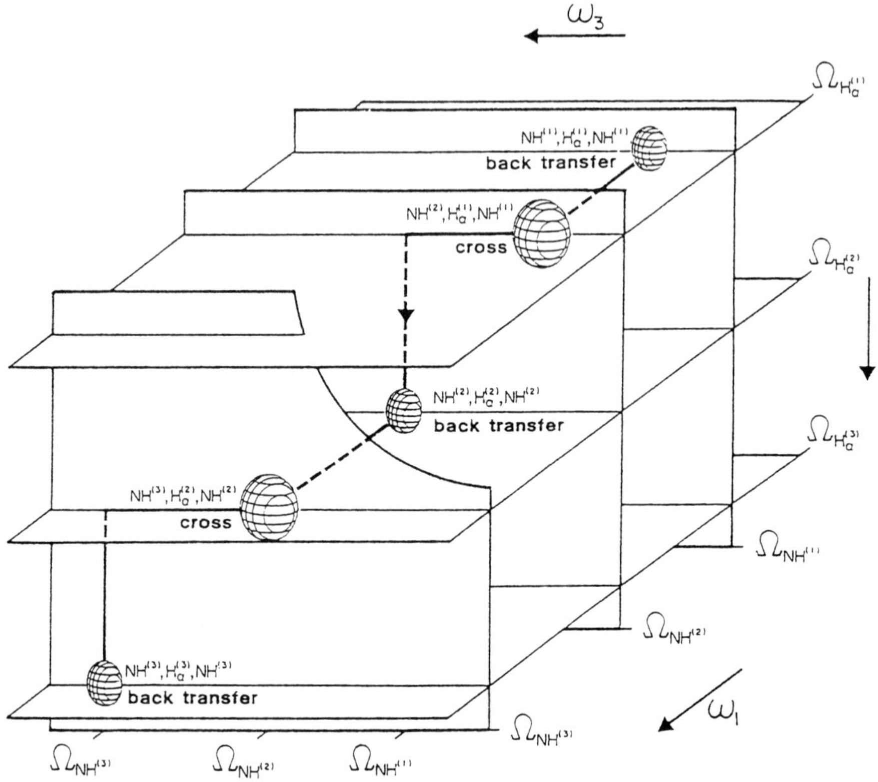

Fig. 11 Schematic representation of the sequential assignment procedure in a selective 3D NOESY-TOCSY experiment. The procedure starts with back-transfer peak NH [1], H_α [1], NH [2]. A linear search parallel to ω_1 leads to the cross-peak NH [2], H_α [1], NH [1]. In a second step, a linear search at coordiante $\omega_1 = \Omega_{NH}$ [2] within the back-transfer plane which is indicated only by dotted lines in the (ω_1, ω_2-planes allows one to reach the next back-transfer peak NH [2], H_α [2], NH [2]). This procedure can be applied repetitively. *Reprinted from Ref. 15.*

$^{15}N_{i+1}$ / NH_{i+1} / $H_{\alpha,i}$ with the $^{15}N_{i+1}$ / NH_{i+1} / $H_{\alpha,i+1}$ involves search on a line (bc in Table 3) whereas the other namely the correlation of the $^{15}N_{i+1}$ / NH_{i+1} / $H_{\alpha,i+1}$ with the $^{15}N_{i+2}$ / NH_{i+2} / $H_{\alpha,i+1}$ (ab in Table 3) involves search in a plane through the $H_{\alpha,i+1}$ resonance. If also ^{13}C labelling is available a true 3D sequencing procedure based on the evaluation of ^{15}N, H-hetero-COSY-TOCSY and hetero-COSY-NOESY as well as ^{13}C, H-hetero-COSY-TOCSY and hetero-COSY-NOESY spectra could be performed. The types of peaks are summarized in Table 4.

Tab. 4

Residue:	i-1				i			
Resonance:	^{15}N	NH	$H\alpha$	$^{13}C\alpha$	^{15}N	NH	$H\alpha$	$^{13}C\alpha$
a	1	2	3					
b		3	2	1				
c			2	1		3		
d			3		1	2		

a Intraresidual cross-peak in ^{15}N, H-COSY-TOCSY or ^{15}N, H-COSY-NOESY.

b Intraresidual cross-peak in ^{13}C, H-COSY-TOCSY or ^{13}C, H-COSY-NOESY.

c Interresidual cross-peak in ^{13}C, H-COSY-TOCSY.

d Interresidual cross-peak in ^{15}N, H-COSY-NOESY.

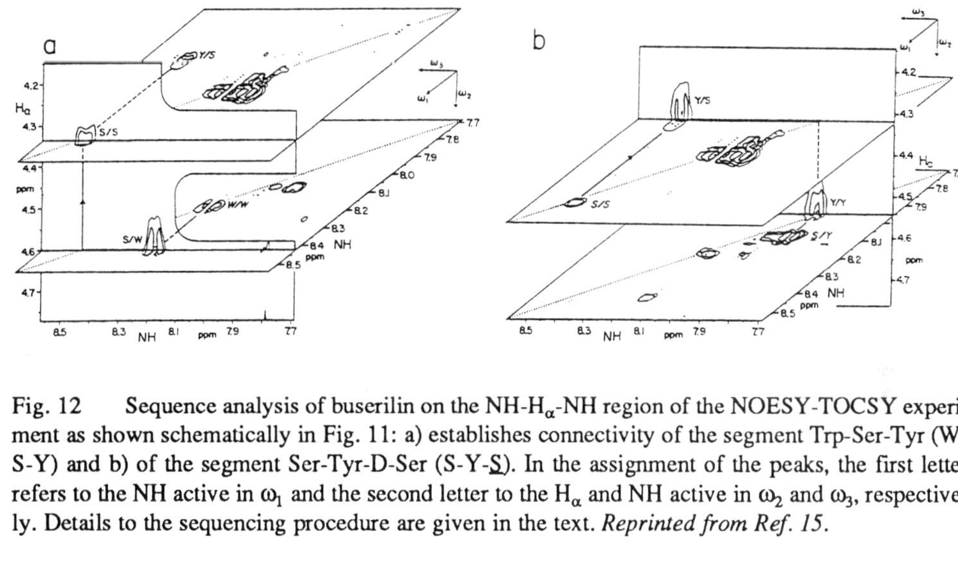

Fig. 12 Sequence analysis of buserilin on the NH-H_α-NH region of the NOESY-TOCSY experiment as shown schematically in Fig. 11: a) establishes connectivity of the segment Trp-Ser-Tyr (W-S-Y) and b) of the segment Ser-Tyr-D-Ser (S-Y-S̲). In the assignment of the peaks, the first letter refers to the NH active in ω_1 and the second letter to the H_α and NH active in ω_2 and ω_3, respectively. Details to the sequencing procedure are given in the text. *Reprinted from Ref. 15.*

3.9 Quantitative Analysis

While assignment procedures require a qualitative interpretation which profits from the increased spectral dispersion of 3D spectra, quantitative data are needed to determine cross-relaxation rates and J-coupling constants, respectively, which provide structural information.

3.9.1 a) Cross-relaxation rates

Cross-relaxation rates that can be measured by the integration of cross-peaks in 2D NOESY or ROESY spectra provide direct distance information. In a 3D spectrum containing a cross-relaxation step in connection with another usually coherent coherence transfer step the cross-peak intensity will be affected by the efficiency of the second transfer step. The most favourable case is then the use of a heteronuclear transfer step via one-bond couplings. Its transfer efficiency ($\sin^2(\pi J\tau)$) (compare to the pulse sequence in Fig. 4f) varies only little due to the small variations of heteronuclear coupling constants. The T_2-dependence ($\exp(-2\tau / T_2)$) is also small, because $T_2 / {}^1J << 1$. For an assumed coupling constant of ${}^1J({}^{15}NH)$ = 90 Hz and a proton linewidth varying between 1 and 16 Hz, the relaxation loss will result in a maximum variation of the apparent internuclear distance of 5% for $\tau = (2J_{NH})^{-1}$.

3.9.2 b) J-coupling constants

The determination of accurate homonuclear or heteronuclear ${}^2J$ or ${}^3J$ coupling constants from 3D spectra is feasible in exceptional cases only as high resolution is required in at least two dimensions, while the third one with reduced resolution can be used for separating overlapping multiplet patterns. For the two high resolution domains ω_2 and ω_3 it is convenient to employ E.COSY-like experiments [32-34] while the first transfer step should produce in-phase magnetization transfer as in TOCSY or in refocussed hetero-COSY.

3.10 Conclusion

3D NMR spectroscopy has a great potential for the NMR spectroscopical analysis of macromolecules. The possibility to unravel complex 2D spectra by the introduction of a third frequency domain is immediately obvious by the use of N^{15}

or C^{13} isotopically enriched macromolecules. Molecular biological methods make them more and more easily available. Heteronuclear correlation experiments which make use of large chemical shift range of ^{15}N or ^{13}C to spread out proton proton transfer spectra can be employed. The heteronuclear correlation sequence has in addition the advantage of almost no loss in sensitivity compared to the conventional 2D experiment and allows the quantitative analysis.

Also proton chemical shifts are used in the third frequency domain. Especially the NOESY-TOCSY experiment should be mentioned. 3D assignment procedures based on homonuclear as well as heteronuclear transfer experiments have been proposed. Qualitative information is most easily extracted from such spectra.

It should be stressed, that quantitative interpretation is more difficult in 3D spectra compared to 2D, because 3D transfer efficiencies depend on the product of two transfer coefficients. Their separate measurement is often impossible. However, heteronuclear transfer spectra are well suited for the quantitative extraction of structural information.

Acknowledgment:
I am grateful to Professor R.R. Ernst in whose laboratory most of the above described work has be done, as well as to Dr. O. W. Sørensen, with whom I experienced a fruitful collaboration.

References

[1] Griesinger, C., Sørensen, O.W. and Ernst, R.R. *J. Magn. Reson.* (1987) *73*, 574.

[2] Ernst, R.R., Bodenhausen, G., Wokaun, A. "Principles of NMR in One and Two Dimensions", Clarendon Press, Oxford (1987).

[3] Wüthrich, K. "NMR of Proteins and Nucleic Acids", Wiley Interscience, New York, (1986).

[4] Aue, W.P., Bartholdi, E. and Ernst, R.R. *J. Chem. Phys.* (1976) *64*, 2229.

[5] Macura, S. and Ernst, R.R. *Mol. Phys.* (1980) *41*, 95.

[6] Braunschweiler, L. and Ernst, R.R. *J. Magn. Reson.* (1983) *53*, 521.

[7] Bothner-By, A., Stephens, R.L., Lee, J., Warren, C.D., Jeanloz, R.W. *J. Am. Chem. Soc.* (1984) *106*, 811.

[8] Müller, L. *J. Am. Chem. Soc.* (1979) *101*, 4481.

[9] Bendall, M.R., Pegg, D.T., Doddrell, D.M. *J. Magn. Reson.* (1983) *52*, 81.

[10] Chandrakumar, N. and Subramanian, S. "Modern Techniques in High Resolution FT-NMR", Springer Verlag, NY. Inc. (1987).

[11] Kessler, H., Gehrke, M., Griesinger, C. *Angew. Chem. Int.* Ed. Engl. (1988), 490-536.

[12] Wagner, G. *J. Magn. Reson.* (1984) *57*, 497.

[13] Griesinger, C., Sørensen, O.W. and Ernst, R.R. *J. Am. Chem. Soc.* (1987) *109*, 7227.

[14] Oschkinat, H., Griesinger, C., Kraulis, P.J., Sørensen, O.W., Ernst, R.R., Gronenborn, A.M., Clore, G.M. *Nature* (1988) *332*, 374.

[15] Griesinger, C., Sørensen, O.W. and Ernst, R.R. *J. Magn. Reson.* (1989) *84*, 14.

[16] Vuister, G.W. and Boelens, R. *J. Magn. Reson.* (1987) *73*, 328.

[17] Vuister, G.W., Boelens, R., Kaptein, R. *J. Magn. Reson.* (1988) *80*, 176.

[18] Fesik, S. and Zuiderweg, R.P. *J. Magn. Reson.* (1988) *78*, 588.

[19] Davies, S.W., Friedrich, J., Freeman, R. *J. Magn. Reson.* (1988) *76*, 555.

[20] Kessler, H., Oschkinat, H., Griesinger, C. and Bermel, W. *J. Magn. Reson.* (1986) *70*, 106.

[21] Cohen, A.D., Freeman, R., McLauchlan, K.A. and Whiffen, D.H. *Mol. Phys.* (1963) *7*, 45.

[22] Nagayama, K. *J. Chem. Phys.* (1979) *72*, 4404.

[23] Eich, G.W., Bodenhausen, G. and Ernst, R.R. *J. Am. Chem. Soc.* (1982) *104*, 3731.

[24] Thayer, A.M., Miller, J.M. and Pines, A. *Chem. Phys. Lett.* (1986) *129*, 55.

[25] Kreis, R., Thomas, A., Studer, W. and Ernst, R.R. *J. Chem. Phys.* (1988) *89*, 6623.

[26] Griffey, R.H., Redfield, A.G. *Q. Rev. Biophys.* (1987) *19*, 51.

[27] Bauer, J., Freeman, R., Frenkiel, T., Keeler, J. and Shaka, A.J. *J. Magn. Reson.* (1985) *58*, 442.

[28] Griesinger, C., Gemperle, C., Sørensen, O.W. and Ernst, R.R. *Mol. Phys.* (1987) *62*, 295.

[29] Warren, W.S. *J. Chem. Phys.* (1984) *81*, 5437.

[30] Loiaza, F., McCoy, M., Hammer, S.L. and Warren, W.S. *J. Magn. Reson.* (1988) *77*, 175.

[31] Wagner, G., Anil Kumar, Wuethrich, K. *Eur. Biochem.* (1981) *114*, 375.

[32] Griesinger, C., Sørensen, O.W. and Ernst, R.R. *J. Am. Chem. Soc.* (1985) *107*, 6394.

[33] Griesinger, C., Sørensen, O.W. and Ernst, R.R. *J. Chem. Phys.* (1986) *85*, 6837.

[34] Griesinger, C., Sørensen, O.W. and Ernst, R.R. *J. Magn. Reson.* (1987) *75*, 474.

[35] Bax, A., Davis, D.G. *J. Magn. Reson.* (1985) *65*, 355.

[36] Griesinger, C., Otting, G., Wüthrich, W. and Ernst, R.R. *J. Am. Chem. Soc.* (1988) *110*, 7870.

[37] Garbow, J.R., Weitekamp, D.P. and Pines, A. *Chem. Phys. Lett.* (1982) *93*, 505.

4. Solution Structure Refinement using Complete Relaxation Matrix Analysis of 2D NOE Experiments: DNA Fragments

Thomas L. James, Brandan Borgias, Anna Maria Bianucci and Ning Zhou

4.1 Introduction

A method for the complete determination of high-resolution molecular structures in *solution* has been a goal of scientists since X-ray diffraction techniques were developed for crystal structure determination. In general, researchers were forced to accept limited structural information due to the techniques available; certainly high-resolution structures such as those derived from X-ray diffraction (XRD) on crystals could not be remotely attained. Methods that yield at least a part of the structural picture include: hydrodynamic techniques, electron diffraction and other scattering techniques, circular dichroism, fluorescence decay, extended X-ray absorption fine structure, and magnetic resonance. The information obtainable from these methods range from rough estimates of the overall shape and dimension of a molecule, to a gauge of the extent of secondary structure (e.g. local conformations properties such as α-helix and β-sheet in the case of proteins). At best it is possible to obtain a measure of some of the distances within the molecule. However, with the possible exception of nuclear magnetic resonance, [1,2] a high-resolution structural model is not yet attainable with any of these methods.

Two recent developments applied in conjunction provide a direct route to molecular structure in non-crystalline phases: 2D NMR and calculational strategies, e.g., the distance geometry (DG) algorithm [3,5], molecular mechanics (MM) [6,7], and molecular dynamics (MD) [7,8]. The essential ideas can be described quite simply. The structure of any molecule can be determined if one can obtain a sufficient number of structural constraints, e.g., internuclear distances and bond torsion angles, to use in conjunction with holonomic constraints of bond lengths, bond angles, and steric limitations. A high-resolution structure requires a large number of such constraints. One can either include (MM,MD) or not include (DG) energetic considerations. Although it will probably not be possible to match the resolution of X-ray crystallography, in principle, the new 2D NMR procedures enable us to obtain many structural constraints, and can lead to high-resolution structures. Fig. 1 shows a proposed scheme which combines experimentally observable structural constraints with various computational

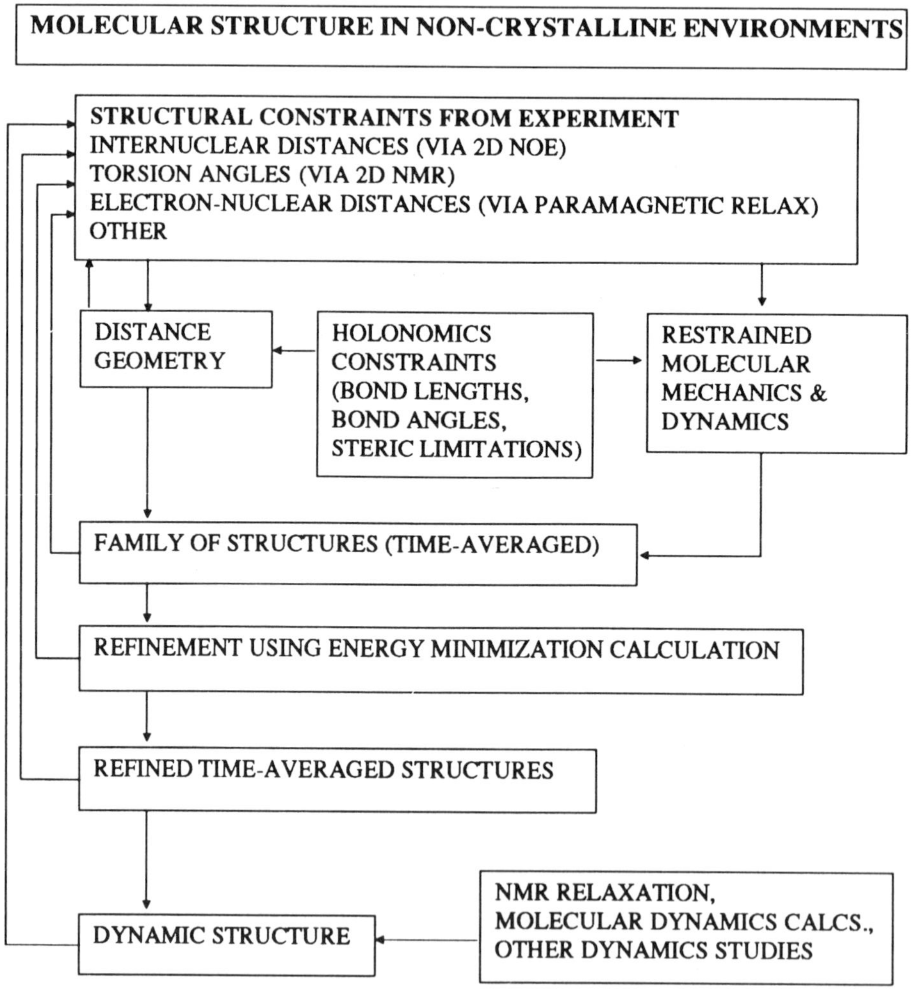

Fig. 1 Scheme for deriving molecular structures in non-crystalline environment using experimental NMR data in conjunction with computational procedures.

strategies to generate structures. Although the scheme in its entirety has not been accomplished, its component parts have been or are being developed.

NMR has experienced an explosion of activity in recent years due largely to the development of 2D NMR. The 2D NMR experiment, first suggested by Jeener [9], is now routinely used in a vast array of specialized experiments for spectrum assignment and structural characterization of nucleic acids and proteins. Key attractions of the 2D NMR methods are that they achieve resolution of crowded spectral regions by mapping the spectrum onto two frequency axes, and they enable simultaneous examination of spin interaction pathways between all affected nuclei. A substantial problem, however, is to assign resonance frequencies to individual nuclei in the biopolymer; even for small proteins, where could be more than 500 protons. However difficult the task, assignment methodologies are being worked out [1,2]. With the advent of 2D NMR techniques, coupled with powerful structure development and refinement techniques such as distance geometry, energy minimisation, and restrained molecular dynamics, detailed structural models of proteins in solution are now becoming achievable. In this paper, we will principally examine the capability of using the two-dimensional nuclear Overhauser effect (2D NOE) experiment, which has been introduced in recent years [10]; it has the potential for providing numerous internuclear distances.

The homonuclear 2D NOE (two dimensional nuclear Overhauser effect) experiment readily yields qualitative structural information. This experiment relies on proton-proton dipolar relaxation to yield a correlation map between protons in close spatial proximity (≤ 5 Å). The intensities of the cross-peaks in the 2D NOE spectrum are related to the distances between the protons and can therefore be used to obtain estimates of those distances. The use of 2D NOE in macromolecular structure determination is now becoming widespread. But, while considerable success has been achieved with this technique, there are certain limitations and precautions that are not always considered and will therefore be listed below.

The 2D NOE experiment was first described (along with the mathematically equivalent experiment for chemical exchange) and given a firm theoretical foundation in Ernst's laboratory [10,11]. We previously described a matrix method for the explicit calculation of the homonuclear 2D NOE intensities expected for a known molecular structure [12], and compared results so obtained for a rigid test molecule (proflavin) in solution with its X-ray crystal structure [13]. The essential concepts described in these seminal papers have been further elaborated by several other laboratories [14,18]. We call this general approach to the calculation and interpretation of 2D NOE intensities CORMA for COmplete RElaxation Matrix Analysis. The critical feature of CORMA is the explicit treatment of the complete relaxation network and specifically spin diffusion in calculating the

2D NOE intensities. An alternative (and widely used) approach assumes that, with short experimental mixing times, a cross-peak intensity correlating a particular proton-proton pair depends on the distance separating that pair of protons alone. This assumption leads to the Isolated Spin Pair Approximation (ISPA). Most analyses of 2D NOE spectra to date have incorporated ISPA in the qualitative or semi-quantitative extraction of interproton distance limits based upon relative cross-peak intensities. Gronenborn and Clore [19] provide a good review of this approach. Later we will demonstrate that the assumption of ISPA is often not valid in practical situations. It might be mentioned that other approximate methods have also been proposed which obviate the worst problems of ISPA [20,21].

The problems facing researchers concerned with DNA structure and protein structure are often of a different nature. In general, we are interested in fairly subtle structural changes in the DNA helix which are sequence-dependent and, consequently, guide protein or drug recognition. These subtle variations demand detailed knowledge of the structure and, therefore, accurate internuclear distance determinations. But one can probably define a protein tertiary structure with moderate accuracy using distance geometry or even restrained molecular dynamics without accurately determining interproton distances. A qualitative assessment of the 2D NOE spectrum is often all that is needed to obtain the information necessary for calculation of a modestly high-resolution protein structure in solution. If (nearly) all proton resonances of a small protein are resolved and assigned and if the protein has a high content of α-helix and β-sheet, the helices and β-sheet structures can be fairly easily defined by the qualitative presence or absence of certain cross-peaks. Then the observation of a (relatively) few "long-range" (in the primary sequence) cross-peaks will serve to orient the secondary structures relative to one another. One can often get a rough idea of the structure immediately, but the use of distance geometry or restrained molecular dynamics will improve the picture and, not coincidentally, lend credence to the structure (or, more properly, family of acceptable structures) obtained. In the most favourable cases, RMS distance deviations of about 1.0 Å can be obtained for the protein backbone. The loop regions and terminal segments are usually the least defined moieties of these protein structures. So, the structures of proteins with low content of common secondary structural elements will be more difficult to pin down. In proteins possessing less common structural features, it may be especially valuable to have more accurate interproton distances for use with the computational techniques. But, more importantly, we will want better defined structures at ligand binding sites (with and without ligand bound). Use of a complete relaxation matrix approach offers the opportunity of determining protein solution structure with greater accuracy and resolution. Protein dynamics can potentially be incorporated into the analysis.

At the very least, a comparison of the experimental 2D NOE spectra with theoretical 2D NOE spectra calculated, using our program CORMA or a similar program, for any nucleic acid or protein structure proposed by distance geometry or restrained molecular dynamics calculations, should be made. So far, no one has reported doing that for proteins. However, we have been doing that for our DNA structural studies for the past few years [22,25]. Recently, we have been examining oligonucleotides containing alternating d(AT) duplexes since (a) such sequences are ubiquitous in the promoter region of genes being recognized by RNA polymerase; (b) it has not been possible to crystallize oligomers with d(TA) segments longer than a tetramer; and (c) at least seven structures have been proposed for alternating d(TA) sequences, largely on the basis of X-ray fiber diffraction data. Examples in this paper will be drawn from that interest.

4.2 Distance Information from 2D NOE Experiments

The basic 2D NOE experiment utilizes the three-pulse sequence: (delay time–$90°– t_1– 90°– \tau_m– 90°– t_2)_n$. The effect of the first 90° pulse is to create transverse magnetization. During delay time t_1 free precession occurs which has the effect of frequency-labelling each of the spins. The second pulse achieves inversion of the spins. During the mixing time τ_m the effect of magnetization exchange due to cross-relaxation is recorded by the spins in the form of population changes. The final 90° observation pulse again generates transverse magnetization for signal detection during t_2. Double Fourier transformation with respect to times t_1 and t_2 results in the 2D NOE spectrum. The effect of cross-relaxation between neighbouring protons during τ_m is to transfer magnetization between them. This results in cross-peak intensities that are roughly inversely proportional to the sixth power of the distance between them. Cross-relaxation during the mixing time τ_m is described by the system of equations [10]:

$$\frac{\partial M}{\partial t} = - RM \tag{1}$$

In eqn. (1), M is the magnetization vector describing the deviation from the thermal equilibrium ($M = M_z – M_0$), and R is the matrix describing the complete dipole-dipole relaxation network. This is essentially an extension of the two-spin equations of Solomon [26]:

$$R_{ii} = 2\,(n_i - 1)\,(W_1^{ii} + W_2^{ii}) + \sum_{j \neq i} n_j\,(W_0^{ij} + 2W_1^{ij} + W_2^{ij}) + R_{1i} \qquad (2a)$$

$$R_{ij} = n_i\,(W_2^{ij} - W_0^{ij}) \qquad (2b)$$

Here n_i is the number of equivalent spins in a group such a methyl rotor, and the zero, single and double transition probabilities W_n^{ij} are given (for isotropic random reorientation of the molecule) by:

$$W_0^{ij} = \frac{q\,\tau_c}{r_{ij}^6} \qquad (3a)$$

$$W_1^{ij} = 1.5\,\frac{q\,\tau_c}{r_{ij}^6}\,\frac{1}{1 + (\omega\,\tau_c)^2} \qquad (3b)$$

$$W_2^{ij} = 6\,\frac{q\,\tau_c}{r_{ij}^6}\,\frac{1}{1 + 4\,(\omega\,\tau_c)^2} \qquad (3c)$$

where $q = 0.1\gamma^4\hbar^2$. The term R_{1i} represents external sources of relaxation such as paramagnetic impurities and is generally ignored. The system of eqns. (1-3) has the solution

$$M(\tau_m) = a(\tau_m)\,M(0) = e^{-R\,\tau_m}\,M(0) \qquad (4)$$

where a is the matrix of mixing coefficients which *are* proportional to the 2D NOE intensities. This matrix of mixing coefficients is what we wish to evaluate. The exponential dependence of the mixing coefficients on the cross-relaxation rates complicates the calculation of intensities (or the distances). Note that the expression for the above rate matrix is actually still an approximation in that it neglects cross-correlation terms between separate pairwise and higher order interactions [18,27]. However, the importance of the cross-correlation terms appears to be small for $\tau_m \leq \tau_m^{opt}$, the optimal mixing time for maximum cross-peak intensity [18]. The expression given above also do not account for second-order effects due to strong scalar coupling [16,28]. Explicit account of J coupling can be included in the analysis of the 2D NOE intensities. Kay et al [16] investigated 2D NOE spectra with cross-relaxation in the presence of strong J coupling. They compared the magnitude of error due to neglect of scalar coupling and found at most a 10% error, except in the case of very strong coupling ($J/\delta \geq 0.5$) and short mixing times (50 ms) where the error could be ≈ 30% [28]. The conclusion is that neglect of J coupling is probably satisfactory in most practical cases where cross-peaks arising from strongly coupled protons are generally not resolvable anyway [28].

Below we will compare a few different methods of analyzing the 2D NOE spectra for internuclear distance and structural content. This analysis will entail use of hypothetical data sets. This use of hypothetical data is necessary, however, since we must know the structure and molecular dynamics exactly in order to understand the effects of any random or systematic error in experimental spectra intensities or the limitations of the different methods being developed to determine structure. We can calculate the theoretical 2D NOE spectrum for the hypothetical structure using any motional model. We can add random noise at any level desired. And we can consider any number of peaks to be overlapping. Furthermore, we can compare the various methods proposed in their abilities to handle realistic spectral limitations. In other words, for a given dynamic structure we can create spectra with various realistic problems. Then we can see how well we are able to deduce the structure using the different methodologies without using our *a priori* knowledge of the structure.

4.2.1 Isolated Spin Pair Approximation (ISPA)

The exact evaluation of intensities (the mixing coefficients a) according to eqn. (4) is hindered by the absence of a general analytical expression for matrix exponentiation. Historically, this had led to the ISPA approach since the exponential can be recast into a Taylor series expansion: (5)

$$a(\tau_m) = e^{-R\tau_m} \approx 1 - R\tau_m + ½R^2\tau_m^2 - \ldots + \frac{(-1)^n}{n!}R^n\tau_m^n + \ldots$$

Truncating the series after the linear term results in a simple approximation for the mixing coefficients that is valid for short ($\tau_m \rightarrow 0$) mixing times:

$$a_{ij}(\tau_m \rightarrow 0) \approx (\delta_{ij} - R_{ij}\tau_m) \tag{6}$$

Incorporation of eqns. (2b), (3a) and (3c) into eqn. (6) gives for $i \neq j$:

$$a_{ij} = n_i \frac{q\,\tau_m}{r_{ij}^6}\left[1 - \frac{6}{1 + 4(\omega\tau_c)^2}\right] \tag{7}$$

The qualitative result is that the intensities are inversely proportional to the sixth power of the distance between the correlated protons.

In some studies, NOE build-up curves are obtained to assess whether or not the short mixing time condition is achieved. Fig. 2 illustrates the validity of ISPA at short mixing times compared to a complete relaxation matrix analysis, calculated with the program CORMA, for the hypothetical three-spin system shown in the figure inset. The curves are calculated using a spectral density appropriate for isotropic motion with a correlation time of 2.4 ns. If the correlation time is larger, corresponding to slower motions, the deviation of the ISPA- and CORMA-calculated curves occurs at a lower τ_m value.

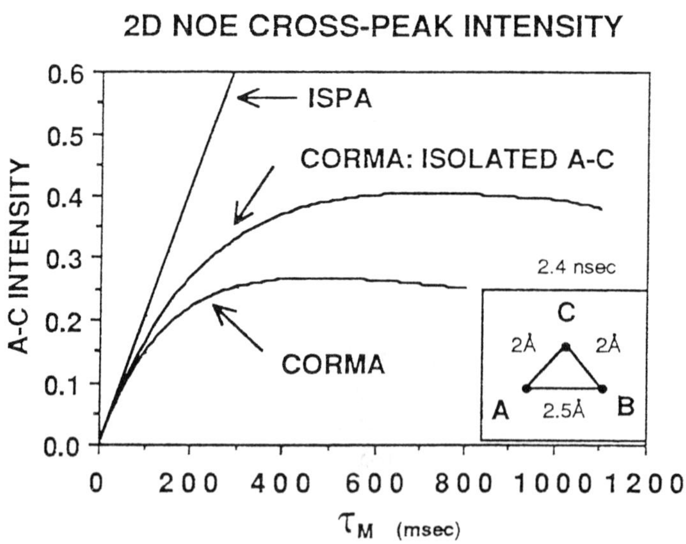

Fig. 2 Cross-peak intensity between two protons in a three-spin system (inset) as a function of mixing time τ_m calculated for an isotropic correlation time of 2.4 ns at 500 MHz. The curves were calculated using the isolated spin pair approximation or the complete relaxation matrix analysis, without the inclusion of the effects of the third proton (see text).

The practical application of ISPA usually goes one step further to eliminate the dependence on correlation time by scaling all the distances with respect to a known reference distance which is assumed to have the same correlation time as the proton-proton pair of interest. Then distances are calculated according to:

$$r_{ij} = r_{ref}\left[\frac{a_{ref}}{a_{ij}}\right]^{1/6} \tag{8}$$

Aside from the errors associated with network relaxation (i.e. failure of the isolated spin pair assumption: $r_{ij} \ll r_{ik}$ and $r_{ij} \ll r_{jk}$ for all k atoms), this protocol references all distances (up to ~ 5.0 Å) to a single short distance that may not be representative of most of the distances determined from the 2D NOE spectrum.

Use of short mixing times definitely limits the signal-to-noise ratio obtainable with cross-peaks in the 2D NOE spectrum. One might also pose the question: how reliable are distances calculated using ISPA? This question was examined by carrying out calculations as follows: (1) Simulate (using CORMA) proton 2D NOE spectra (500 MHz) for a DNA duplex octamer in B form, i.e. without assumptions of ISPA, for an effective isotropic correlation time of 4 ns at a few mixing times. (2) Assuming these simulated 2D NOE spectra were experimentally obtained, calculate interproton distances using ISPA with the following guidelines for oligonucleotides [19]: (a) the correlation times for sugar-sugar, sugar-methyl, and sugar-base (with the exception of H1'-base) inter-proton vectors are the same as the intranucleotide H2'-H2" vector (reference distance 1.77 Å); (b) the correlation times for base-base, base-methyl and H1'-base inter-proton vectors are the same as the intranucleotide H5-H6 vector (reference distance 2.46 Å). (3) Compare the internuclear distances calculated via ISPA with the distances from the model B-DNA.

Fig. 3 shows that comparison using the C2' geminal protons for reference. Clearly, analysis with ISPA introduces a systematic error, calculated distances are shorter than true distances. As the distances become larger, both the systematic deviation and the scatter increase. The amount of error between true and calculated distances depends on the exact environment of the proton pair. If a longer reference distance is used, the "data points" in the above figure effectively move up relative to the line of unit slope, so that shorter distances will be overestimated while the longer distances will still be underestimated (but not as badly). It should be noted that the shortest mixing time illustrated, 50 ms, is actually shorter than that which is sometimes employed. But the deviation between true and calculated distances clearly becomes larger with longer mixing time. *Deviations can be expected to be worse with proteins, which possess higher proton density, and with slower motions.*

If approximate ISPA distances are desired, it is still important to restrict the analysis to data collected in the time-domain where the approximation is valid. The question of how short a mixing time is needed for the truncation of eqn. (5) to be reasonably accurate has not generally been answered, although mixing times typically used are frequently in the 50-100 ms range. Is this short enough? That question can be answered by running through the series expansion term-by-term for a representative case.

Intensities were calculated according to eqn. (5) for protons in B-DNA assuming a correlation time of 4 ns (at 500 MHz, $\omega\tau_c = 12.6$), and mixing times of 10, 50, 100, 150 and 200 ms. The results are given in Table 1. At 10 ms, intensities were in error by only 5-15%. However, intensities for distances between 1.8 Å and 2.3 Å, were found to be overestimated by 20-90% at 50 ms. The intensities for these cross-peaks are expected to be dominated by direct relaxation. Intensities for longer distances, where spin-diffusion is expected to make a significant contribution, were found to be underestimated by 50-70%. Moreover, simply increasing the number of terms in the expansion proved to be a poor route to obtain intensities. We found that the series converges slowly, requiring three terms at

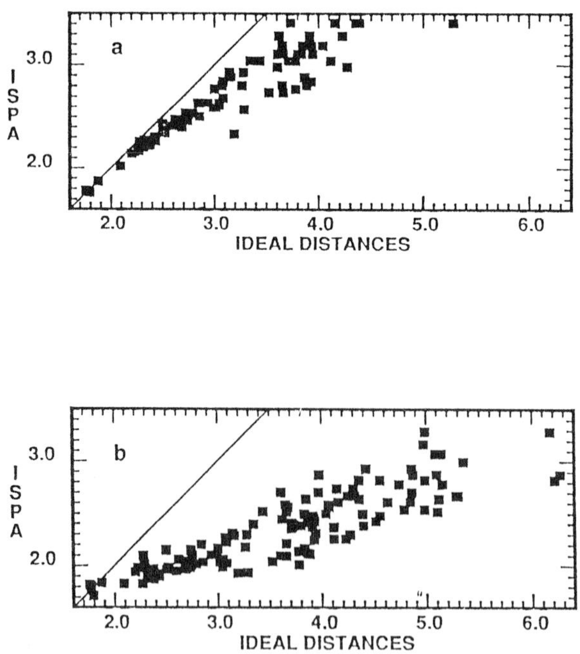

Fig. 3 Comparison of distances calculated according to the isolated spin pair approximation (ISPA) with actual distances. The ISPA distances were calculated from ideal intensities generated by CORMA for d(GGTATACC) in energy-minimized B-DNA conformation assuming a correlation time of 4 ns and a mixing time of a) 50 ms, and b) 250 ms.

10 ms and as many as 13 terms to converge for a 200 ms experiment with a correlation time of 4 ns (see Table 1).

Tab. 1 Convergence of Series Expansion for 2D NOE Intensity Calculation

		Error[c] in Single Term Approximation	
τ_m[a]	N[b]	r_{ij} = 1.77 to 2.27 Å	r_{ij} = 3.60 to 7.15 Å
10	3	+5 to 15%	-0 to -30%
50	4	+40 to 90%	-45 to -70%
100	7	+60 to 205%	-50 to -80%
150	10	+90 to 410%	-60 to -90%
200	13	+130 to 570%	-60 to -90%

[a] Mixing time for calculation in milliseconds. Correlation time is assumed to be 4 ns, and the operating frequency is 500 MHz ($\omega\tau_c$=12.6).

[b] Number of terms in series expansion required to achieve less than a 5% deviation in intensities averaged over all calculated intensities.

[c] Error is defined as: $(I_I-I_N)/I_N$, where I_I is the single term intensity, and I_N is the intensity calculated after N terms. Reported in these columns are the ranges of errors encountered for several distances as indicated. Intensities for short distances (≤ 3.0 Å) are typically overestimated by the single term approximation. For longer distances the intensities are typically underestimated. The range of errors quoted here actually correspond to errors for specific distances of 1.77, 2.22, 2.27, 3.60, 4.22, 5.53 and 7.15 Å.

The trade-off between performing matrix diagonalization (*vide infra*) and series summation occurs at about three terms. Hence only for very short mixing times would a series expansion ever be useful.

In summary, there are two problems with using ISPA. The requisite short mixing times give 2D NOE spectra with smaller signal-to-noise. Whenever at least one proton approaches either of the "isolated pair" (or both) at a distance less than the distance between the pair, the approximation breaks down for practical values of τ_m for molecules with an effective correlation time greater than a nanosecond. In fact, there are on average, 3.4 neighbouring protons within a 3.0 Å radius of each proton in B-DNA. But for DNA, the reference distances typically used are the geminal H2'-H2" distance of 1.77 Å, and the H5-H6 distance of 2.46 Å in cytosine. Moreover, this protocol references all distances (up to 5 Å) to a single short distance that is not really representative of the majority of distances determined from the 2D NOE spectrum.

The implication of this for using ISPA-derived distances with distance geometry is that the bounds, in particular the upper bound, may need to be relaxed more than has been the case in calculations reported. There are also implications for restrained molecular dynamics calculations. If the distances are systematically underestimated, the protein may never be able to get to the vicinity of the global minimum during the MD simulation.

4.2.2 Complete Relaxation Matrix Analysis (CORMA)

CORMA Method. A more expeditious method of calculating intensities is to take advantage of linear algebra and the simplifications which arise from working with the characteristic eigenvalues and eigenvectors of a matrix. The rate matrix R can be represented by a product of matrices: $R = \chi\lambda\chi^T$ where χ is the unitary matrix of orthonormal eigenvectors ($\chi^{-1} = \chi^T$), and λ is the diagonal matrix of eigenvalues. The utility of making this transformation is that, since λ is diagonal, the series expansion for its exponential (and consequently that of the mixing coefficient matrix) collapses:

$$a = 1 - \chi\lambda\chi^T \tau_m + \tfrac{1}{2}\, \chi\lambda\chi^T \chi\lambda\chi^T \tau_m^2 - \ldots \qquad (9)$$

$$a = \chi e^{-\lambda\tau_m} \chi^T \qquad (10)$$

This calculation allows one to readily calculate all the cross-peak intensities for a proposed structural model. Then comparison between calculated and measured intensities allows a determination as to the validity of the model structure. We have developed a program for performing this calculation, named CORMA, which is available upon request from the authors [12]. Fig. 4 is a schematic representation of intensities (generated by the program CORMA) for the aromatic-H1' region of $[d(GGTATACC)]_2$ in a B-D-B conformation in comparison with experimentally obtained intensities. An alternative representation, using contours and true chemical shift axes, as generated by the program LINSHA [29], is depicted in Fig. 5.

The complete relaxation matrix approach (a) is accurate, (b) can accommodate any size spin system (computer size limitations only), (c) is not limited to any range of mixing times, (d) incorporates spin diffusion naturally, and (e) can utilize any molecular motion model. We have carried out a number of theoretical calculations as well as experimental studies. In summary [12,13,30]: (a) distances up to 6 Å (and possibly 7 Å) with an accuracy of 10% could be attainable with knowledge of individual relaxation times. (b) detailed knowledge of the molecular motions is not required for this distance accuracy; it is generally sufficient for a single *effective* isotropic correlation time to be used in the spectral density expression for any nucleus. One can iteratively fit the simple 2D NOE spectra of small molecules [13], but that was initially not possible for larger molecules. Over the course of our studies, the sophistication of analysis developed from evaluation of a few models on the basis of selected intensities to

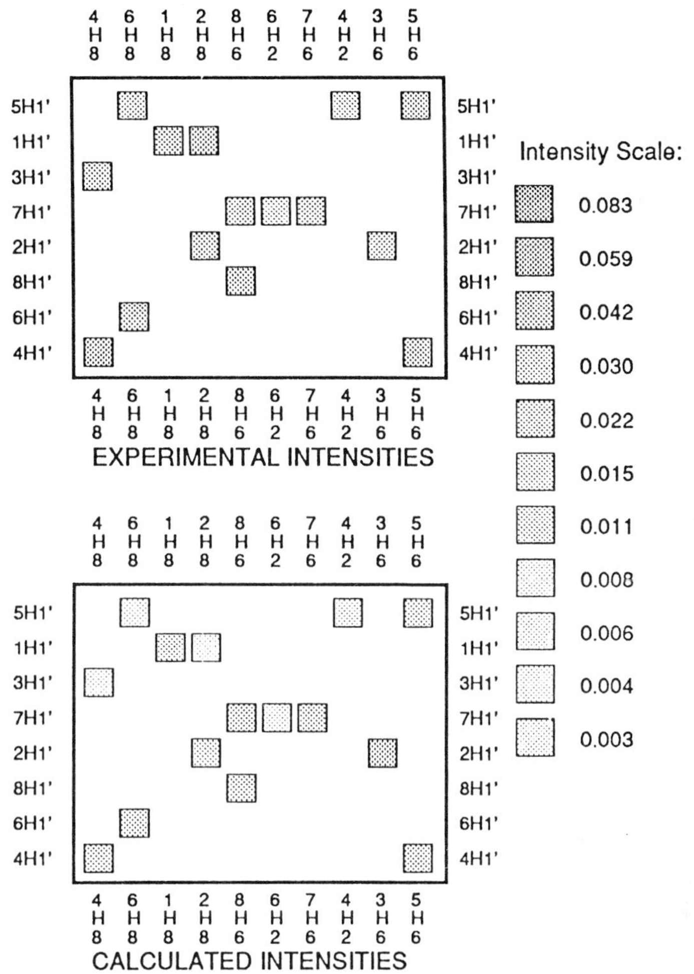

Fig. 4 Schematic intensities for the aromatic-H1' region in $[d(GGTATACC)]_2$ as determined by CORMA. The atoms were ordered according to chemical shift. Experimental intensities were obtained at 500 MHz with a mixing time of 250 ms. Calculated intensities assume a correlation time of 4 ns, and B-D-B conformation. Only these peaks which were unambiguously assigned and measured with high confidence are shown in these plots.

a more detailed analysis of many closely related models using all well-resolved intensities [22,25].

Application to DNA Fragments. The first oligonucleotide we studied was $[d(GGAATTCC)]_2$ [22,31] in which the central six base pairs comprise the *Eco*RI endonuclease recognition site. These six base pairs are also identical to those in the dodecamer $[d(CGCGAATTCGCG)]_2$, for which a crystal structure has been reported [32]. The 2D NOE intensities involving the non-exchangeable base protons were compared with CORMA intensities calculated for both regular

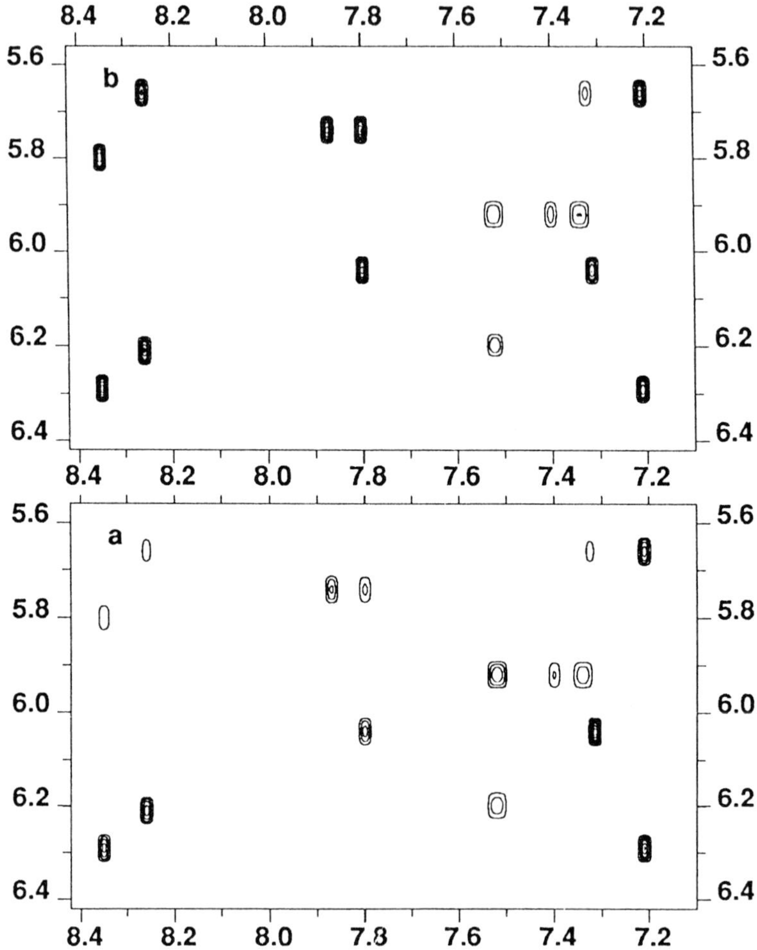

Fig. 5 Contour plots of the intensities shown in Fig. 4 as generated by software developed by Dr. Hans Widmer. CORMA calculated intensities are shown in a), experimental intensities are shown in b). Line shapes for both plots are calculated assuming a natural linewidth of 5 Hz, and acquisition times of 0.15 s in both t_1 and t_2, and 45° phase-shifted sine-squared apodization was applied in both domains. Coupling constants were assigned to model the splitting in the experimentally obtained spectrum (not shown). Peak assignments are as shown in Fig. 4 and correspond to: A4-H8 8.35, A6-H8 8.26, G1-H8 7.87, G2-H8 7.80, C8-H6 7.52, A6-H2 7.40, C7-H6 7.34, A4-H2 7.33, T3-H6 7.26, T5-H6 7.21, A4-H1' 6.29, A6-H1' 6.21, C8-H1' 6.20, G2-H1' 6.04, C7-H1' 5.92, T3-H1' 5.80, G1-H1' 5.74, T5-H1' 5.66 ppm. Displaying the intensities in this manner results in greater apparent contrast than the simply grey scale representation of Fig. 4. However, the validity of this display is subject to the selection of appropriate coupling constants and the assumption that all peaks have the same natural linewidth.

B-DNA and for a structure obtained as a result of energy minimization using the program AMBER (Assisted Model Building with Energy Refinement) [6,7]. Neither structure accurately reproduced all peak intensities, but many features of the experimental data were represented by both the regular B form and the energy-minimized structure. Overall, the energy-minimized structure manifested intensities which best matched the experimental intensities. It was noted that the effective correlation times involving the base protons were found to lie in the range 1.7-2.7 ns with the A H8 protons exhibiting the longer correlation times. While most of the intensities were best fit with the energy-minimized structure, the T(5)-H6 protons were better fit by the regular B structure. These observations suggested a sequence-dependent structural change at the purine/pyrimidine junction in the oligomer. The plausibility of this is supported by crystallographic results: when *Eco*RI endonuclease was co-crystallized with $[d(TCGCGAATTCGCG)]_2$ [33], a kink was found at the GAA-TTC junction. It was suggested that this kinked conformation, while clearly stabilized by binding of the DNA to the protein, must also exist transiently for the isolated oligonucleotide alone in solution.

Another study involved the duplex $[d(GGTATACC)]_2$, containing the TATA segment which is commonly found in the promoter region of genes and binds to RNA polymerase preparatory to transcription. Twenty 2D NOE cross-peaks at each of four mixing times (100, 175, 250 and 400 ms) were compared with CORMA-generated intensities for regular B-DNA and for the analogous energy-minimized structure [23]. However, neither trial structure adequately modelled the solution structure; the GG and CC regions were well represented by regular B-DNA , but the TATA moiety exhibited large discrepancies with both models.

Interest in the alternative d(TA) structure suggested a study on the decamer duplex $[d(AT)_5]_2$ [24]. The structure of the alternating d(TA) sequence has been subject to some controversy. On the basis of NMR studies, it has been proposed to have a normal B-DNA structure [34,35] or a left-handed B conformation in solution [36]. And seven models have been proposed for its structure in the solid state. Comparison of experimental intensities with CORMA-calculated intensities assuming a 7 ns correlation time for structures A, B, alternating B, left-handed B, C, D, wrinkled D and Z (derived from $[d(CGCGCG)]_2$), showed most favourable fits for B, D and wrinkled D-DNA structures, with the wrinkled D form providing the best fit to experimental data.

Molecular mechanics (AMBER) calculations also indicated that the wD structure was the best model for $[d(AT)_5]_2$ in moderate salt solution. Interesting structural features of wD DNA include (a) 8 basepairs per helical turn; (b) a much narrower minor groove, which has the consequence of creating a hydration tunnel; (c) significant cross-strand hydrophobic interactions; and (d) alternating torsion angle values at TA and AT steps [24].

Subsequently, complete re-analysis of the 2D NOE spectra of $[d(GGTATACC)]_2$ from our previous study [23] was carried out. As mentioned above, our earlier study was unable to fit the TATA moiety, although the GC base pairs assumed the B-DNA structure. Considering the apparent preference for the wD structure for $[d(AT)_5]_2$, we built a model for $[d(GGTATACC)]_2$ in which the -TATA- segment exists in the wD form and GG/CC moieties in regular B form (hereafter referred to as BDB model). Seventeen Na^+ ions and nine water molecules were included in this BDB model for subsequent energy-minimization based on the following structural features we found in the $[d(AT)_5]_2$ study [24]. The BDB model was built using wD DNA coordinates for the GG and CC moieties. Protons were added using standard bond-lengths and bond-angles. To improve atom contacts at the structural junction, the -GT- -CA- part was then subjected to energy-minimization (with AMBER, *vide infra*) with the rest of the molecule fixed. In short, the BDB model thus constructed was able to yield theoretical 2D NOE spectra in good agreement with the set of four experimental spectra. The 2D NOE intensities corresponding to these models were then compared with the experimental intensities previously generated [23]. Additionally, the 2D NOE intensities corresponding to four other models: A, crystalline A [37], B, and energy-minimized B (with the inclusion of counter ions and water hydration) were also considered. In this study [25], all non-overlapping cross-peaks were quantified from all four mixing times (100, 175, 250 and 400 ms), yielding a total of 542 cross-peak intensities. The energy-minimized BDB structure (Fig. 6) gave the best fit to the experimental intensities and was also most

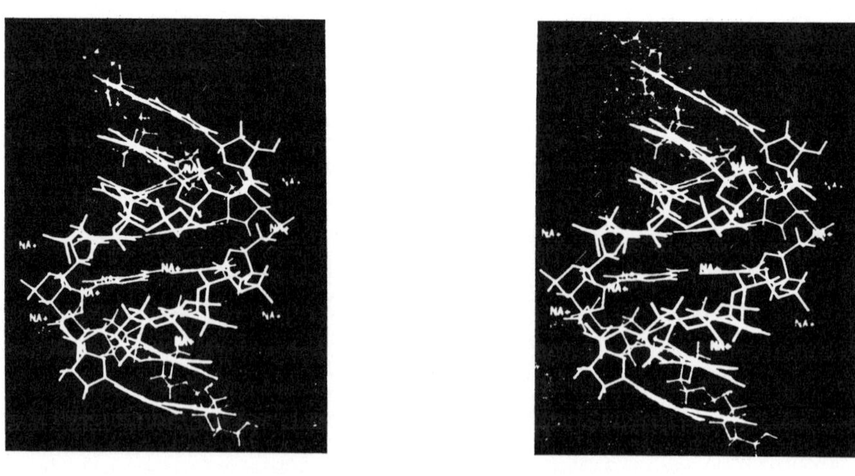

Fig. 6 Stereo diagram for the energy-minimized BDB model of $[d(GGTATACC)]_2$, shown with counter ions $[NA^+]$. The minor groove is orientated towards the reader. Water molecules are omitted for clarity.

favourable from an energetic standpoint. Like the wrinkled-D $[d(AT)_5]_2$ decamer, it also contains a hydration tunnel. However, in this structure, the tunnel is sealed off by the $[d(GG)]_2$ and $[d(CC)]_2$ caps which are in a predominantly B-DNA conformation. The energy-minimized structure generated in this study also gave a reasonable fit, as opposed to the energy minimized B structure in the prior study [23]. This is presumably due to the inclusion of counter ions and hydration water during the minimization.

4.2.2.1 Direct Calculation of Distances (DIRECT).

The major drawbacks of the trial-and-error approach (*vide supra*) is that it is limited by the choice of structural models. It is capable only of discriminating between the proposed structures and of indicating regions of good (or bad) fit between the model and true solution structure. There are two possible ways to circumvent this limitation. One is the direct determination of distances without making the isolated spin pair approximation - essentially using the reverse of CORMA. The other is to automatically refine the structure based on the 2D NOE intensities.

An obvious solution to the problem of limited (and biased) trial structures is to apply the same computational techniques used in CORMA to the direct calculation of proton-proton distances from the experimental intensities. This approach has been discussed elsewhere [14,30]. Rearrangement of eqn. (4) gives:

$$\frac{-\ln\left[\dfrac{a(\tau_m)}{a(0)}\right]}{\tau_m} = R \qquad (11)$$

So, assuming isotropic tumbling and using eqns. (2) and (3), distances can be calculated directly from the rate matrix via the experimental intensities. For relatively small systems, in which most of the major peaks (in particular the diagonals) can be resolved and accurately estimated, this is clearly the ideal method of distance determination. However, low resolution (peak overlap) and low (i.e. generally realistic) signal-to-noise hamper the accurate estimation of distances [30].

We have tested the performance of the direct calculation of distances (referred to as the DIRECT method) in the presence of several different types of data errors: random noise, cross-peak overlap, and diagonal peak overlap. CORMA was used to calculate intensities at mixing times of 50 ms and 250 ms for a single strand of d(GGTATACC) in the energy-minimized BDB conformation. Intensities were calculated with a noise level of ± 0.003 (which is experimentally typical), or to a precision of five significant figures with no noise. The effects of cross-peak overlap were modelled by selecting for analysis only those intensities which correspond to well-resolved and assignable peaks in an experimental spectrum. The effects of diagonal peak overlap were considered by assigning an average (over all the calculated diagonal intensities) to each of the diagonals. The distances calculated by this approach compare favourably with ISPA distances. The results of one test are presented in Fig. 7. Both methods accurately assign the

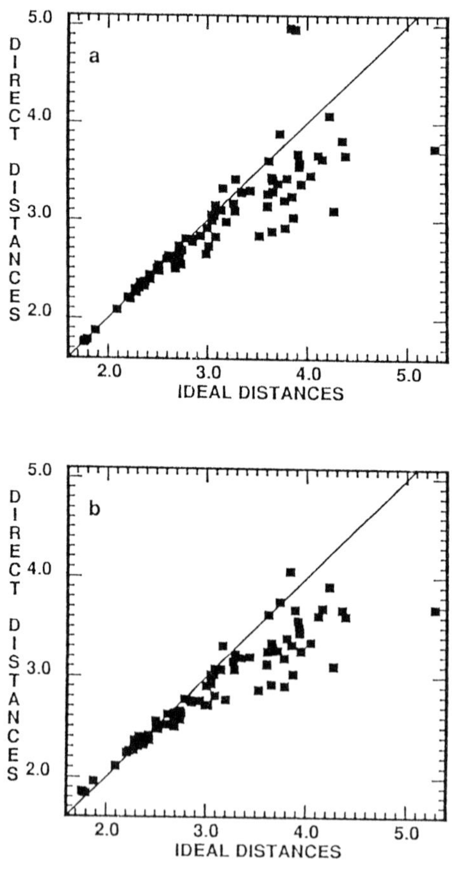

Fig. 7a Comparison of distances calculated directly from the diagonalized scaled intensity matrix (DIRECT method) with ideal distances. DIRECT distances were calculated according to eqns. (2-3) and (11). Intensities were calculated by CORMA for d(GGTATACC) in energy-minimized B-DNA conformation assuming an isotropic correlation time of 4 ns. a) 50 ms mixing time, all calculated intensities ≥ 0.003, b) 50 ms mixing time, intensities selected to correspond to experimentally resolved peaks.

distances for very short interactions, and short mixing times. However, even with predicted distances of only 3-4 Å, the accuracy of the distance calculations is already questionable. ISPA severely underestimates all the distances due to the neglect of spin diffusion and relaxation from nearby protons, while the DIRECT calculation suffers from the neglect of significant relaxation pathways when all the cross-peaks or diagonals cannot be properly assigned or measured. Still, the DIRECT calculation is a substantial improvement over ISPA and should find much use.

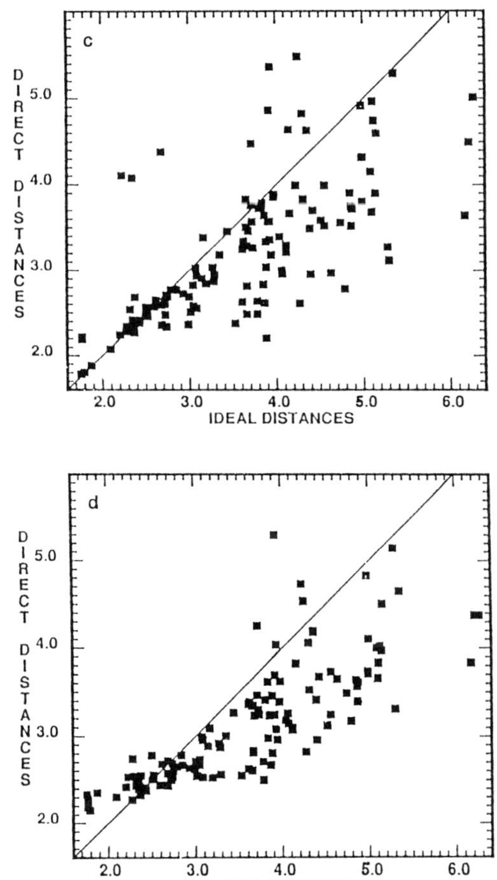

Fig. 7b Comparison of distances calculated directly from the diagonalized scaled intensity matrix (DIRECT method) with ideal distances. DIRECT distances were calculated according to eqns. (2-3) and (11). Intensities were calculated by CORMA for d(GGTATACC) in energy-minimized B-DNA conformation assuming an isotropic correlation time of 4 ns. c) 250 ms mixing time, all calculated intensities ≥ 0.003, and d) 250 ms mixing time, intensities selected to correspond to experimentally resolved peaks.

4.2.2.2 Structural Refinement using COMATOSE.

The comparatively poor prospects for biopolymer structure determination by the DIRECT approach led us to pursue the less direct and more time consuming approach of iterative least-squares refinement of structure based on the 2D NOE intensities. With the increasing availability of more powerful computers, it is becoming feasible to refine the structures of modest-sized biopolymers (i.e. those with *ca.* 600 protons) using their 2D NOE spectra. We have begun developing the program COMATOSE (COmplex Matrix Analysis Torsion Optimized StructurE) for the refinement of molecular structure based on 2D NOE intensities [30]. The goal is to optimize a trial structure while minimizing the error between calculated and observed intensities. We have incorporated CORMA into a program with a Marquardt-Levenberg least-squares optimizer. Consequently, we investigated the effects of experimental errors in peak intensities on the refinement process.

The use of Cartesian coordinates to define the position of protons provided too many parameters for refinement to proceed properly. To reduce the number, one can take advantage of holonomic structural constraints. Several alternative algorithms using different variable parameters were tried, but the most promising one makes use of torsion angles and group orientation parameters rather than the full set of coordinates. For nucleic acids, the following parameters are defined for each nucleotide: χ, the glycosidic bond torsion angle; γ, the C4'-C5' bond torsion angle; sugar pucker parameters P and θ_{max} [38], group orientation parameters, α', β', γ', x, y, and z. The sugar pucker parameters P θ_{max} allow fewer parameters since the ring torsions are correlated and constrained by the condition of ring closure, and are used to define each of the furanose ring torsions [38]. Each nucleotide is described as a freely floating entity, its location described by the group orientation parameters, which is held in place only by the constraints imposed by the fit between calculated and observed internucleotide NOE intensities. The internucleotide torsion angles are not specifically included in the description of the structure, although constraints such as the O5'-O3' angle could be included during refinement. This yields ten structural parameters for each nucleotide.

Refinement against torsion angles as variable parameters has the advantage that the number of parameters is considerably reduced from the alternative choice of Cartesian coordinates. Furthermore, incorporation of ancillary constraints such as bond lengths, bond angles and atom connectivities, as well as others (*vide infra*) is easily accomplished.

We have found COMATOSE to work reasonably well for DNA fragments. Fig. 8 depicts the results from one study. This study was for a 250 ms mixing time with experimentally realistic random noise, providing nearly optimal spectral S/N; the other methods (ISPA, DIRECT) compared were not very good at such a long mixing time. Consequently, it is feasible that sufficient cross-peak intensity can be measured that distances out to 7 Å can be determined as illustrated. Not only does COMATOSE improve the distances, but the spectral density can be refined as well. For the figure shown, the original hypothetical model for the study had an isotropic correlation time of 4.0 ns. The COMATOSE calculations though were initiated with a correlation time of 10 ns, i.e., as a rough guess one might choose in the absence of any other information. COMATOSE refinement yielded a correlation time of 4.2 ns, simultaneously yielding the distances plotted. This level of performance has been found in many cases now.

It is obvious that an initial structure is necessary for refinement via COMATOSE. In the case of DNA fragments, it is reasonable to start with standard B form DNA and refine that structure using the experimental 2D NOE spectra. In the case of proteins, it may be appropriate to obtain an initial structure using ISPA (or, better, DIRECT) via analysis of the distances with either the methods of distance geometry or restrained molecular dynamics. COMATOSE could then be used to refine that structure. In short, the COMATOSE refinement process is much more cpu time consuming than simpler methods for estimating the distances from 2D NOE intensities such as ISPA or the DIRECT method. However, it does not suffer from the systematic errors associated with such methods. No information is needed from the diagonal peak intensities (which are typically overlapping and difficult to measure), and overlapping of tiny cross-peaks pose no problem (other than loss of potential additional information) since these intensities are simply not included in the calculation of the error to be minimized (alternatively the error in the *sum* of the calculated intensities corresponding to these overlapped peaks can be added to the error). Only well-resolved peaks need be considered, while their intensities are calculated taking into account the complete relaxation matrix analysis (CORMA). Major limitations of the COMATOSE approach are (a) the neglect of energetic and steric constraints during the structure refinement process and (b) ability of COMATOSE to find a local minimum rather than a global minimum. These occasionally allow conformations to be generated which satisfy the distance constraints imposed by the NOE intensities, but which are otherwise unacceptable.

The 2D NOE data from $[d(GGTATACC)]_2$ were used with COMATOSE to see if the structure could be improved beyond that of the energy-minimized BDB model which we had previously [25] found satisfactory from our model-building. As indicated in Fig. 9, optimization with COMATOSE did improve the fit of the resultant structure's theoretical 2D NOE spectra to the experimental spectra. We also could utilize the optimized distances derived from COMATOSE as restraints using a molecular mechanics (AMBER) force field containing a harmonic NOE distance restraint term. We have used such restrained AMBER calculations on $[d(GGTATACC)]_2$, which had already been subjected to COMATOSE optimization (A.M. Bianucci, B. Borgias and T.L. James, unpublished results). The results (simplified) are summarized in Fig. 9. An advantage of our approach may be apparent from this scheme. That is, at any stage of structure refinement, we can see how well we are doing by calculating the theoretical 2D NOE spectra and comparing with the experimental spectra.

Acknowledgements

The work described here was supported by the National Institute of Health (Grants GM 39247 and CA 27343) and the National Science Foundation (Grant PCM 84-04198)

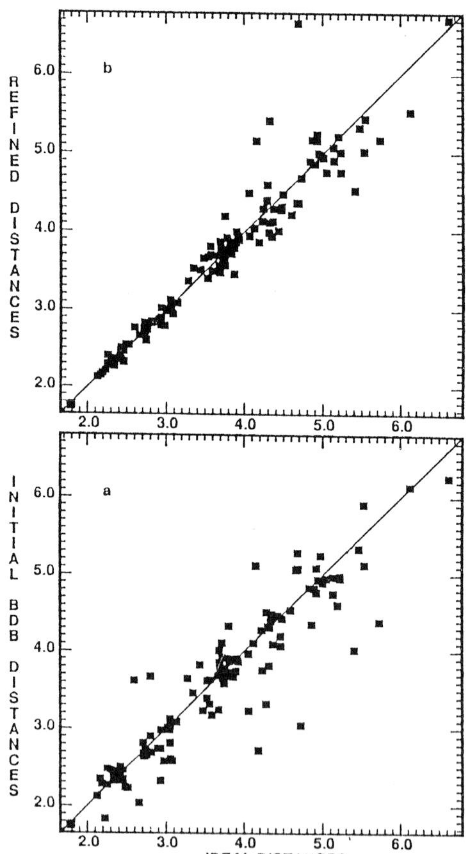

Fig. 8 Comparison of distances obtained from COMATOSE refinement with ideal distances. Intensities were calculated by CORMA for $[d(GGTATACC)]_2$ in energy-minimized BDB·DNA conformation assuming an isotropic correlation time of 4 ns and a mixing time of 250 ms. The starting conformation was in BDB-DNA conformation prior to energy minimization. a) Prior to COMATOSE refinement. b) After COMATOSE refinement.

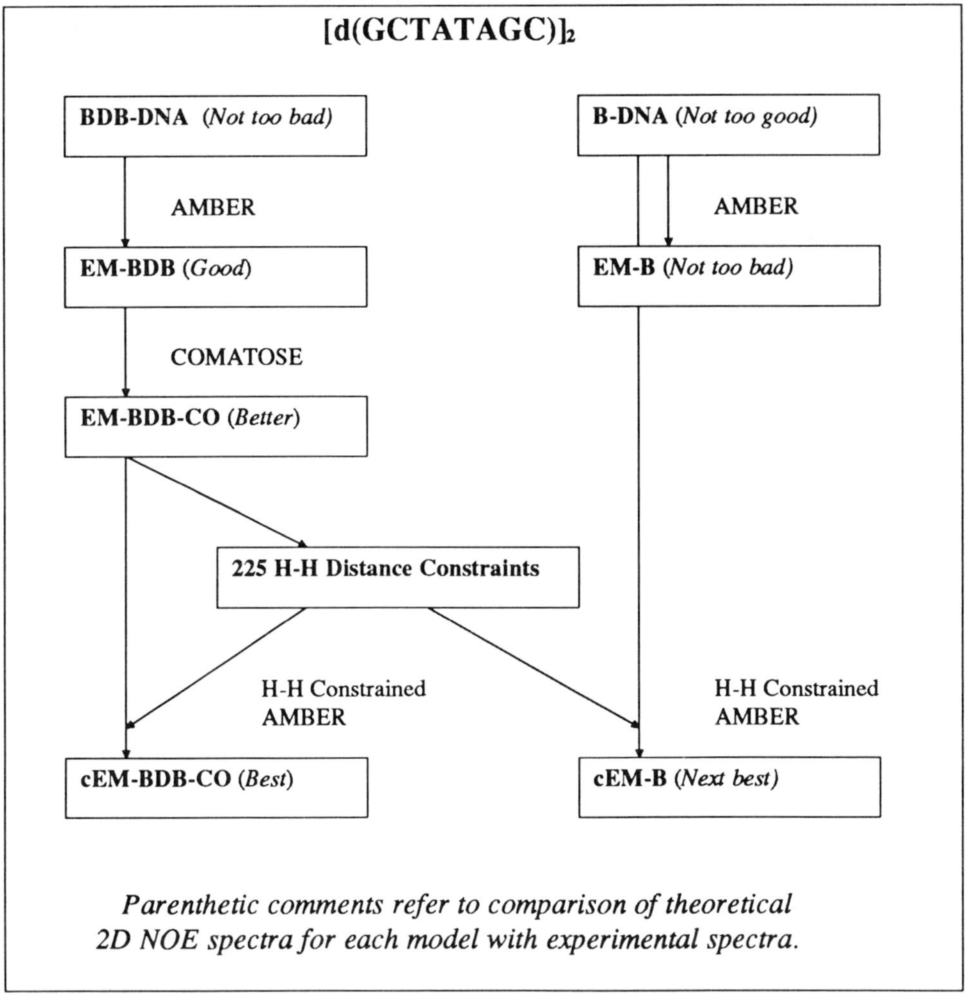

Fig. 9 Refinement of $[d(GGTATACC)]_2$ structure via molecular energy refinement (AMBER), without and with COMATOSE-derived NOE-distance restraints, and optimization of theoretic 2D NOE spectra against experimental 2D NOE spectra using COMATOSE. EM and cEM prefixes to the model names denote energy-minimized, without and with NOE-distance constraints, respectively, and the CO suffix denotes COMATOSE-optimized.

References

[1] Wüthrich, K. "NMR Proteins and Nucleic Acids", Wiley, New York, (1986).

[2] Oppenheimer, N.J. and James, T.L.Eds. "NMR in Enzymology" Academic Press, New York, (1988) in press.

[3] Havel, T.F., Kuntz, I.D. and Crippen, G.M. *Bull. Math. Biol.* (1983) *45*, 665-720.

[4] Braun, W. and Go, N. *J. Mol. Biol.* (1985) *186*, 611-626.

[5] Havel, T.G. and Wüthrich, K. *J. Mol. Biol.* (1985) *182*, 281-294.

[6] Weiner, P.K. and Kollman, P.A. *J. Comp. Chem.* (1981) *2*, 287-303.

[7] Singh, U.C., Weiner, P.K., Caldwell, J. and Kollman, P.A., AMBER 3.0, University of California, San Francisco, (1986).

[8] Karplus, M. and McCammon, J.A. *Nature*, (1979) *277*, 578.

[9] Jeener, J. "," AMPERE International Summer School Basko Polje, Yugoslavia, (1971).

[10] Macura, S. and Ernst, R.R., *Mol. Phys.* (1980) *41*, 95-117.

[11] Jeener, J., Meier, B.H., Backmann, P. and Ernst, R.R. *J. Chem. Phys.* (1979) *371*, 4546-4553.

[12] Keepers, J.W. and James, T.L. *J. Magn. Reson.* (1984) *57*, 404-426.

[13] Young, G.B. and James, T.L. *J. Am. Chem. Soc.* (1984) *106*, 7986-7988.

[14] Olejniczak, E.T., Gampe, Jr.R.T. and Fesik, S.W. *J. Magn. Reson.* (1986) *67*, 28-41.

[15] Macura, S., Farmer, B.T. and Brown, L.R. *J. Magn. Reson.* (1986) *70*, 493-499.

[16] Kay, L.E., Scarsdale, J.N., Hare, D.R. and Prestegard, J.H. (1986) *J. Magn. Reson.* *68*, 515-525.

[17] Massefski, Jr.W. and Bolton, P.H. *J. Magn. Reson.* (1985) *65*, 526-530.

[18] Bull, T.E. *J. Magn. Reson.* (1987) *72*, 397-413.

[19] Gronenborn, A.M. and Clore, G.M. pp. 1-32 in *Progress in Nuclear Magnetic Resonance Spectroscopy*, ed. L.H. Sutcliffe, Pergammon Press, Oxford, (1985).

[20] Eaton, H.L. and Anderson, N.H. *J. Magn. Reson.* (1987) *74*, 212-225.

[21] Lefevre, J-F., Lane, A.N. and Jardetzky, O. *Biochemistry* (1987) *26*, 5076-5090.

[22] Broido, M.S., James, T.L., Zon, G. and Keepers, J.W. *Eur. J. Biochem.* (1985) *150*, 117-128.

[23] Jamin, N., James, T.L. and Zon, G. *Eur. J. Biochem.* (1985) *152*, 157-166.

[24] Suzuki, E.-I., Pattabiraman, N., Zon, G. and James, T.L. *Biochemistry* (1986) *25*, 6854-6865.

[25] Zhou, N., Bianucci, A.M., Pattabiraman, N. and James, T.L. *Biochemistry* (1987) *26*, 7905-7913.

[26] Solomon, I. *Phys. Rev.* (1955) *99*, 559.

[27] Werbelow, L. and Grant, D.M. *Adv. Magn. Reson.* (1978) *9*, 1989.

[28] Kay, L.E. Holak, T.A., Johnson, B.A., Armitage, I.M., Prestegard, J.H. *J. Am. Chem. Soc.* (1986) *108*, 4242.

[29] Widmer, H. and Wüthrich, K. *J. Magn. Reson.* (1987) *74*, 316-336.

[30] Borgias, B.A. and James, T.L. *J. Magn. Reson.* (1988) in press.

[31] Broido, M.S., Zon, G. and James, T.L. *Biochem. Biophys. Res. Commun.* (1984) *119*, 663-670.

[32] Dickerson, R.E. and Drew, H.R. *J. Mol. Biol.* (1981) *149*, 761-786.

[33] Frederick, C.A., Grable, J., Melia, M., Samudzi, C., Jen-Jacobson, L., Wang, B.-C., Greene, P., Boyer, H.W. and Rosenberg, J.M. *Nature* (1984) *309*, 327-331.

[34] Assa-Munt, N. and Kearns, D.R. *Biochemistry* (1984) *23*, 791-796.

[35] Borah, B., Cohen, J.S. and Bax, A. *Biopolymers* (1985) *24*, 747-765.

[36] Gupta, G., Sarma, M.H., Dhingra, M.M., Sarma, R.H., Rajagopalan, M. and Sasisekharan, V. *J. Biomolec. Struct. Dyn.* (1983) *1*, 395-416.

[37] Shakked, A., Rabinovich, D., Kennard, O., Cruse, W.B.T., Salisbury, S.A. and Viswamitra, M.A. *J. Mol. Biol.* (1983) *166*, 183-201.

[38] Altona, C. and Sundaralingam, M. *J. Amer. Chem. Soc.* (1972) *94*, 8295

5. NMR Studies of Proteins, Nucleic Acids and their Interactions.

Robert Kaptein, Rolf Boelens and Thea M.G. Koning

5.1 Introduction

For several decades X-ray crystallography was the only technique that could yield three-dimensional structures of biomolecules in atomic detail. This situation has changed recently, since NMR spectroscopy has also been developed into a tool for structure determination [1,2]. An advantage of the NMR approach is that it does not rely on a crystalline state of the biomolecules and that their structures are therefore determined in solution, more close to the natural environment. A drawback is the relatively small size of the molecules; so far, structures have been determined of proteins of molecular weight less than 10,000 daltons.

NMR structures are primarily based on a set of short proton proton distances obtained from the Nuclear Overhauser Effect (NOE) [3]. The origin of the NOE is dipolar cross-relaxation between protons. Because of the weak proton magnetic moment and the r^{-6} distance dependence of the effect, NOE's can only be measured between protons at relatively short distances (< 5 Å). Using suitable calibration procedures the NOE's can be translated into constraints on proton proton distances. Several computational methods exist now for structure determination using these distance constraints, such as, for instance, distance geometry [4,5] and restrained molecular dynamics [6,7,8]. As the NOE is a spin-relaxation phenomenon it depends upon the dynamic behaviour of the molecule in solution. Therefore, the structure and the dynamics of a biomolecule as seen by NMR are intimately connected. Indeed, NMR is unique in providing information on the dynamics of molecules as has been realized since a long time.

In this article we shall discuss the methods available for structure determination of biomolecules by NMR. A recent improvement in the proton proton distance determination based on a relaxation matrix description is presented. This so-called iterative relaxation matrix approach (IRMA) can be integrated in a structure determination procedure [9,10]. As an example of a structural problem studied by NMR the work of our laboratory on protein-DNA recognition will be reviewed. The solution conformation of the DNA-binding domain (headpiece) of the *E.coli lac* repressor protein was determined solely on the basis of NMR data [7,11]. Recently, using a similar methodology the structure of a complex of *lac* headpiece with a *lac* operator fragment (its DNA recognition site) could also be determined [12,13].

5.2 Biomolecular Structures from NMR

The protocol that we have used for determination of the three-dimensional structure of *lac* repressor headpiece is shown in Scheme 1.

1	Assign ^{1}H resonances
2	Determine proton proton distance constraints and dihedral angle constraints from NOE's and J-couplings, respectively.
3	Calculate family of structures using geometric constraints only (experimental constraints plus covalent structure) using, for instance, distance geometry (DG) and distance bounds driven dynamics (DDD).
4	Refine these structures using geometric and potential energy functions, for instance, with restrained energy minimization (REM) and restrained molecular dynamics (RMD).

Scheme 1

The first two steps, consisting of ^{1}H resonance assignment and determination of distance and dihedral angle constraints, is common to all procedures. Steps 3 and 4 are suitable to address questions such as how unique are the structures obtained, how well do they satisfy the experimentally derived constraints and how reasonable they are from the point of view of energetics. In the following we shall discuss various steps of Scheme 1 in some detail.

5.2.1 ^{1}H resonance assignments

A necessary requirement for the structural analysis of a protein is the assignment of the great majority of its proton resonances. For small proteins (MW < 10,000) that do not aggregate at millimolar concentrations this can be accomplished using a combination of various 2D NMR experiments. The procedure for the so-called sequential assignment of protein ^{1}H NMR spectra is extensively described by Wüthrich [2]. Briefly, two main classes of 2D experiments are used. In the first off-diagonal cross-peaks arise only between protons connected through J-coupling networks, with COSY (correlated spectroscopy) as the prime example. Another very useful experiment in this category is TOCSY (total correlation spectroscopy [14]) or 2D HOHAHA (Homonuclear Hartmann-Hahn [15]). In these spectra patterns of cross-peaks can be traced between pairs of J-coupled protons as in COSY or between several protons within an amino acid chain as in HOHAHA.

In the second class of 2D NMR experiments cross-peaks connect protons that are spatially in close proximity (distance < 5 Å). The 2D NOE or NOESY experiment [16] and its rotating frame counterpart ROESY [17,18] fall into this category. A 2D NOE spectrum is recorded in a three-pulse experiment [16].

$$90^\circ - t_1 - 90^\circ - t_m - 90^\circ - t_2 \text{ (acq.)}$$

In this sequence 90° stands for a 90° radio frequency pulse, t_1 and t_2 are the variable times which after double Fourier transformation yield the ω_1 and ω_2 frequency domains of a 2D spectrum; t_m is a fixed time, which allows exchange of magnetization between nuclei. The origin of the NOE effect is dipolar cross-relaxation, which depends on fluctuations in the orientation and length of the vectors connecting pairs of nuclei. In a rigid molecule these vectors have fixed lengths and reorient by the tumbling of the molecule as a whole. In that case cross-relaxation rates are proportional to r^{-6} and therefore have a very strong distance dependence. As an example, a 2D NOE spectrum of *lac* repressor headpiece in D_2O is shown in Fig. 1. All off-diagonal intensity in this spectrum corresponds to short distances between non-exchangeable protons.

The sequential assignment usually starts with a search for cross-peak patterns belonging to the spin-systems or types of amino acids. These are then connected through cross-peaks in a 2D NOE spectrum between neighbouring amino acids in the polypeptide chain.

Useful short distances involving backbone protons that are manifested in 2D NOE spectra are those between C_α, C_β and amide protons of one residue and the amide proton of the next residue ($d_{\alpha N}$, $d_{\beta N}$ and d_{NN}, respectively). Often the sequential assignment procedure is redundant so that many internal checks are possible. This makes the assignment unambiguous. However, it does not lead to stereospecific assignments for diastereotopically related protons such as those of methylene groups or methyl groups of valine and leucine. Sometimes stereospecific assignments of these protons is possible using a combination of vicinal J-coupling and NOE information [19,20].

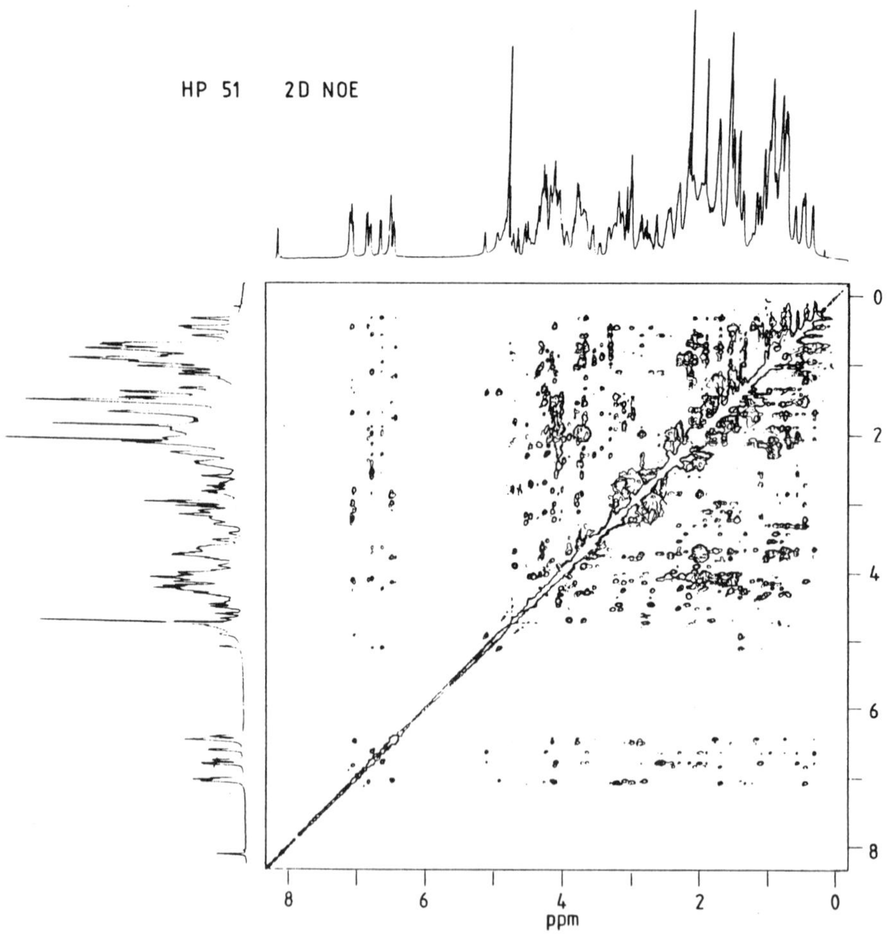

Fig. 1 2D NOE spectrum of headpiece 51 represented as a contour plot. The spectrum was recorded at 500 MHz of a 5 mM solution of headpiece in D_2O. The mixing time t_m was 100 ms. Off-diagonal cross-peaks indicate short distances (< 4 Å) between non-exchangeable protons. *Reprinted from Ref. 54.*

5.2.2 Distance and Dihedral Angle Constraints

Proton proton distance constraints are most conveniently derived from cross-peak intensities in 2D NOE spectra. The initial build-up rate of these cross-peaks in spectra taken at short mixing times is proportional to the cross-relaxation rate σ_{ij} between protons i and j. Therefore, these cross-relaxation rates can be measured either from a single 2D NOE spectrum taken with a sufficiently short mixing time, or, more accurately, from a build-up series recorded with various mixing times. For a rigid isotropically tumbling molecule σ_{ij} is simply related to the distance d_{ij} and the correlation time τ_c:

$$\sigma_{ij} \propto \tau_c \, d_{ij}^{-6} \tag{1}$$

Therefore, using a known calibration distance d_{cal} the proton proton distances follow from the relation

$$d_{ij} = d_{cal} \, (\sigma_{cal} / \sigma_{ij})^{1/6} \tag{2}$$

In practice eqn. (1) and (2) are only approximately valid. There are two main problems associated with accurate determination of proton proton distances. The first is that of indirect magnetization transfer or "spin-diffusion". In reality the NOE cross-peaks are the result of multispin relaxation and only in the limit of extremely short mixing times (where the signal-to-noise ratio is poor) the two-spin approximation of eqn. (1) is valid.

As discussed below the effect of spin-diffusion can be calculated and procedures based on a full relaxation matrix treatment are being developed to solve this problem [21,9]. Secondly, proteins are not rigid bodies and intramolecular mobility leads to non-linear averaging of distances and to different effective correlation times for different interproton vectors in the molecule. Internal motions occur over a wide range of time-scales and only the fast fluctuations (up to a few hundred pico seconds) can presently be simulated by molecular dynamics calculations. The slower motions are difficult to handle and therefore constitute the most serious source of error in distance determination from NOE's.

For this reason, the approach that is usually taken is that of translating the NOE information into distance ranges (e.g. 2-3 Å, 2-4 Å, 2-5 Å for strong, medium and weak NOE's respectively) rather than attempting to obtain precise distances. Alternatively, as was done in the case of *lac* headpiece [7] a single cut-off distance was used at 3.5 Å (corresponding to a range of 2-3.5 Å).

In case stereospecific assignments of methylene and methyl groups are not known the distances involving these protons have to refer to pseudo-atoms and a corresponding correction term has to be added allowing for the maximum possible error. For instance, for a methylene group the pseudo-atom position is defined at the geometric mean of the CH_2 proton positions and a correction of 0.9 Å is added to the distance constraints involving these protons.

The absence of NOE's between assigned protons also contains useful information as it means that the distance between the protons is larger than ca. 4-5 Å. This non-NOE information has been used in the work on the *lac* headpiece structure [22]. Great care should be taken, however, in interpreting the absence of NOE's, since it may also result from local mobility.

In favourable cases dihedral angle constraints can be obtained from three-bond J-couplings. These can be obtained from the fine structure of cross-peaks in COSY-like spectra recorded with high digital resolution [2]. An important example is the three-bond coupling $^3J_{HN\alpha}$ between amide and Cα protons, which gives a measure of the backbone torsion angle ϕ. For helical regions $^3J_{HN\alpha}$ is small (ca. 4 Hz) while for extended chain conformations such as in β-sheets it takes large values (9-10 Hz). Usually the large J-couplings (8-10 Hz) are the most useful source of information, because J-couplings smaller than the line-width (typically 5 Hz or larger) cannot be reliably measured due to cancellation effects in antiphase multiplets [23]. Furthermore, the interpretation of the larger coupling constants in terms of dihedral angles is less ambiguous.

5.2.3 Structure Calculations Based on Geometric Constraints (Distance-geometry, Distance Bounds Driven Dynamics)

The metric matrix distance geometry (DG) algorithm [4,24,26] was known well before protein structure determination by NMR became possible. Thus far, it is the only method that does not rely on some starting conformation and is therefore free from operator bias. The DG procedure amounts to the following. First upper and lower bound matrices U and L are set up for all atom-atom distances of the molecule. Some of the elements u_{ij} and l_{ij} follow from standard bond lengths and bond angles of the covalent structure, and from experimentally found distance ranges from NOE's and J-coupling constants. A bound smoothening procedure using triangle inequalities extends the constraints to all elements of U and L. Then, a distance matrix D is set up with distances chosen randomly between upper and lower bounds, $l_{ij} \leq d_{ij} \leq u_{ij}$. The so called "embedding" algorithm then finds a 3D structure corresponding to the distances d_{ij}. This structure must be optimized using an error function consisting of chirality constraints (chiral centres sometimes come out the wrong way) and a distance constraint error function.

This forces the amino acid side chain to adopt the correct chirality and the distances to satisfy the upper and lower bound criteria, although usually the DG structures still contain a number of violations of the distance bounds. By repeating the DG calculations several times the random step of choosing the distance matrix D between upper and lower bounds allows different structures to be obtained and therefore it can be judged how uniquely the structure is determined by the constraints.

Another method, also termed distance geometry but using a quite different mathematical procedure, was suggested by Braun & Go [5]. Here, the protein conformation is calculated by minimizing a distance constraint error function. Special features of the method are that dihedral angles are used as independent variables rather than Cartesian coordinates and that it uses a variable target function, first satisfying local constraints (between amino acids nearby in the polypeptide chain), while at a later stage long range constraints are included. Usually one starts with various initial conformations obtained by taking random values for the dihedral angles. A comparison between a metric matrix distance geometry algorithm DISGEO [27] and the variable target function algorithm DISMAN [5] has shown that the efficiency and convergence properties of both methods are rather similar [28].

Although the distance geometry method does not need starting structures and is therefore not subject to operator bias, this does not mean that it samples the allowed conformational space (consistent with the bounds) in a truly random fashion. In fact, it was noted by Havel and Wüthrich [26] in model calculations on bovine pancreatic trypsin inhibitor, that the DG structures tend to be somewhat more expanded than the crystal structure, from which the constraints had been derived. We have also noticed this effect in our work on *lac* headpiece [29]. In regions that are relatively unconstrained the DG algorithm tends to produce extended conformations for the backbone and side chains (*vide infra*). It was further noted that in restrained molecular dynamics refinement a larger variation of conformations was found in regions with few NOE's in spite of the fact that the structures now have to satisfy criteria of low potential energy as well. This led to the idea of improving on the sampling properties of the DG procedure by adding a simplified MD calculation with only geometric constraints as the driving potential. In this so-called Distance Bounds Driven Dynamics (DDD) algorithm [30] Newton's equations of motion

$$m_i \ddot{r}_i = F_i \tag{3}$$

are solved with the forces given by

$$F_i = -\frac{\partial V}{\partial r_i} \tag{4}$$

where the potential function now does not contain any energy terms but is taken proportional to the DG error function

$$V = K_{dc}\left[\sum_{d_{ij}>u_{ij}} (d_{ij}^2 - u_{ij}^2)^2 + \sum_{l_{ij}>d_{ij}} (l_{ij}^2 - d_{ij}^2)^2\right] \tag{5}$$

The time-step for integration of Newton's eqn. (3) should be chosen in accordance with the magnitude of the "force costant" K_{dc} (taken somewhat arbitrarily as 10,000 $kJmole^{-1}nm^{-4}$). Of course, the entirely non-physical nature of the potential V means that the calculations cannot be interpreted as a simulation of a physical process. A DDD run using a DG structure as the starting conformation has the effect of "shaking up" the DG structure and thereby improving the sampling of conformation space, which is especially important in regions with few constraints.

5.2.4 Structure Refinement Including Energy Terms (Restrained Energy Minimization and Molecular Dynamics)

The quality of the protein structures based on geometric constraints can be improved by taking energy considerations into account. For instance, in DG structures amino acid side chains often adopt eclipsed conformations, while in alkyl chains the energy of the staggered conformation is 10 to 15 $kJmole^{-1}$ lower. Also, hydrogen bonds and salt bridges may not be formed unless they are specifically introduced as constraints. In restrained molecular dynamics (RMD) refinement structures are optimized simultaneously with respect to a potential energy function and to a set of experimental constraints. Of course the success of this method now depends to a large extent on the quality of the force field used. It is therefore important to realize the limitations and approximate nature of this force field, especially when the calculations do not include solvent molecules.

In RMD calculations eqns. (3) and (4) are integrated with the potential energy function given by

$$V = V_{bond} + V_{angle} + V_{dihedr} + V_{vdW} + V_{Coulomb} + V_{dc} \tag{6}$$

The first two terms tend to keep bond lengths and bond angles at their equilibrium values. V_{dihedr} is a sinusoidal potential describing rotations about bonds; for V_{vdW} (the van der Waals interaction) usually a Lennard-Jones potential is taken, and $V_{Coulomb}$ describes the electrostatic interactions. The extra distance constraint term V_{dc} distinguished RMD from more conventional MD simulations. It has the effect of pulling protons within the distance $d_{ij}{}^0$ (or the distance range) in accordance with the NOE observations. Usually a harmonic pseudo-potential is chosen

$$V_{dc} = \frac{1}{2} K_{dc} \left(d_{ij} - d_{ij}^0\right)^2 \tag{7}$$

Other forms of V_{dc} are a half-harmonic potential [7] and pseudo-potentials that contain a linear part at long distances and repulsive terms for non-NOE's [22]. Similarly, sinusoidal terms describing dihedral angle constraints from J-coupling constants can be included [22].

A restrained dynamics calculation is usually preceded by restrained energy minimization (REM) using steepest descent or conjugate gradient methods to bring the energy down to an acceptable level. REM using the same potential energy function (6) usually changes the conformation only slightly and cannot take it out of local minima. By contrast, RMD is able to overcome barriers of the order of kT because of the kinetic energy in the system and therefore has a much larger radius of convergence. RMD works as an efficient minimizer, since excess potential energy, converted to kinetic energy, is drained off by coupling the system to a thermal bath of constant temperature. It has been suggested [31] that the RMD procedure could be used to obtain folded protein structures starting, for instance, from a fully extended polypeptide chain. Although apparently this was successful for model calculations on crambin, [31] it is our experience that this procedure does not work in general and is certainly not cost effective in terms of best use of computer time. In our view RMD should be considered as a structure refinement tool using DG or DDD structures as starting conformations.

5.3 Iterative Relaxation Matrix Approach (IRMA)

So far the problem of indirect magnetization transfer or "spin-diffusion" in 2D NOE spectra has usually been circumvented by using short mixing times of, say, 50 ms. The disadvantages of this are, first that the signal to noise ratio of the spectra is poor and artefacts due to t_1 noise and zero quantum coherences are relatively large. Secondly, especially NOE cross-peaks involving protons at rather long distances (ca. 4 Å) may still contain an appreciable (but unknown) amount of spin-diffusion. These factors limit the precision of the proton proton distances obtainable from 2D NOE spectra.

It is possible to calculate the effect of spin-diffusion by solving the set of coupled differential equations known as the generalized Bloch equations [32] that govern the multi-spin relaxation problem of all the protons present in a biomolecule. This can be done either by diagonalization of the relaxation matrix [21] or by stepwise integration of the coupled equations [33,34]. However, in order to do this one needs a structural model from which inter-proton distances are measured and, in addition, one has to make assumptions on the motional behaviour of the molecule. Therefore, this approach has thus far been used only to discriminate between a number of model structures for oligonucleotides [35,36].

On the other hand it is known that the relaxation matrix can be calculated by diagonalization of the matrix of NOE intensities and thus information on proton proton distances can be obtained directly. This method was originally suggested for exchange spectroscopy [37,38] and was further elaborated for 2D NOE by Olejniczak et al., [39]. However, although direct back transformation of the NOE (or exchange) matrix may work for small systems, it does not in the case of the more crowded spectra of biomolecules, because these spectra are usually incompletely assigned. Furthermore, the strong diagonal intensities present a problem.

We have recently proposed an iterative procedure that combines elements of the previous approaches, but avoids the above mentioned problems [9,10]. We shall first give a brief description of the underlying relaxation theory and then describe the procedure in more detail.

5.3.1 Theory

Multispin relaxation in a biomolecule in solution can be approximately described by the generalized Bloch equations. In this approach the time development of the matrix of normalized cross-peak intensities A in a 2D NOE spectrum is described by the set of coupled differential equations [32,21].

$$\frac{dA}{dt} = -R\,[A-1]$$

which can be formally solved for a mixing time τ_m as

$$A = \exp[-R\,\tau_m] \tag{8}$$

where R is the cross-relaxation matrix with

$$R = \begin{bmatrix} \rho_1 & \sigma_{12} & \cdot & \cdot & \cdot & \sigma_{1n} \\ \sigma_{21} & \rho_2 & \cdot & \cdot & \cdot & \sigma_{2n} \\ \cdot & \cdot & \cdot & \cdot & \cdot & \cdot \\ \cdot & \cdot & \cdot & \cdot & \cdot & \cdot \\ \cdot & \cdot & \cdot & \cdot & \cdot & \cdot \\ \cdot & \cdot & \cdot & \cdot & \cdot & \cdot \\ \sigma_{n1} & \sigma_{n2} & \cdot & \cdot & \cdot & \rho_n \end{bmatrix} \tag{9}$$

The relaxation rate constants in the dipolar cross-relaxation matrix for a rigid molecule with N protons are defined by

$$\rho_i = K \sum_{\substack{j=1 \\ j\neq i}}^{N} \left(\frac{1}{r_{ij}^{6}}\right) \left[\,6\,J_2(\omega) + 3\,J_1(\omega) + J_0(\omega)\,\right] \tag{10}$$

$$\sigma_{ij} = K \left(\frac{1}{r_{ij}^{6}}\right) \left[\,6J_2(\omega) - J_0(\omega)\,\right] \tag{11}$$

where r_{ij} is the distance between protons i and j and $K = 0.1 \cdot \gamma^4 \cdot \hbar^2$. For isotropic tumbling of the molecule with a correlation time τ_c the spectral densities $J_n(\omega)$ take the simple form

$$J_n(\omega) = \frac{\tau_c}{1 + n^2 \omega^2 \tau_c^2} \tag{12}$$

where ω is the Larmor frequency of the protons. The exponential matrix $\exp[-R\tau_m]$ can be expanded in a power series

$$\exp[-R\,\tau_m] = 1 - R\,\tau_m + 0.1R^2\,\tau_m^2 \ldots \tag{13}$$

For sufficiently short mixing times only the first two terms in eqn. (13) contribute appreciably and the NOE intensity a_{ij} becomes $\sigma_{ij} \cdot \tau_m$ and builds up linearly with the mixing time.
Alternatively, the matrix equations can be solved as

$$A = X \exp[-\Lambda\,\tau_m]\,X^{-1} \tag{14}$$

where Λ is the diagonal eigenvalue matrix obtained after diagonalisation of R

$$X^{-1}\,R\,X = \Lambda \tag{15}$$

Therefore, given a molecular model the NOE matrix can be calculated from the relaxation matrix for each mixing time τ_m of a 2D NOE experiment and the build-up of NOE intensities in a time series can be obtained.

The reverse procedure is also possible [37,38,39]. When the complete NOE matrix A is known, the relaxation matrix R can be obtained after diagonalisation of the NOE matrix.

$$X^{-1}\,A\,X = D = \exp[-\Lambda\,\tau_m] \tag{16}$$

$$R = -X\left[\frac{\ln D}{\tau_m}\right]X^{-1} \tag{17}$$

Thus, theoretically the matrix R can be obtained from one 2D NOE experiment taken at a suitably chosen mixing time. In practice averaging over a series of τ_m values is more accurate [9].

5.3.2 The IRMA Cycle

The procedure is summarized as a flow diagram in Fig. 2. It starts with an initial structural model, possibly obtained from a set of approximate distance constraints. In the case of oligonucleotides we used the canonical A- and B-DNA structures for this purpose. For this starting structure and a choice for the motional behaviour of the molecule the relaxation matrix is set up and the NOE matrix is calculated. The off-diagonal elements of this matrix for which experimental NOE's are available, are then replaced by the measured ones. The combined NOE matrix is diagonalised to obtain a relaxation matrix with cross-relaxation elements now including the effect of spin-diffusion. After averaging a set of matrices for a

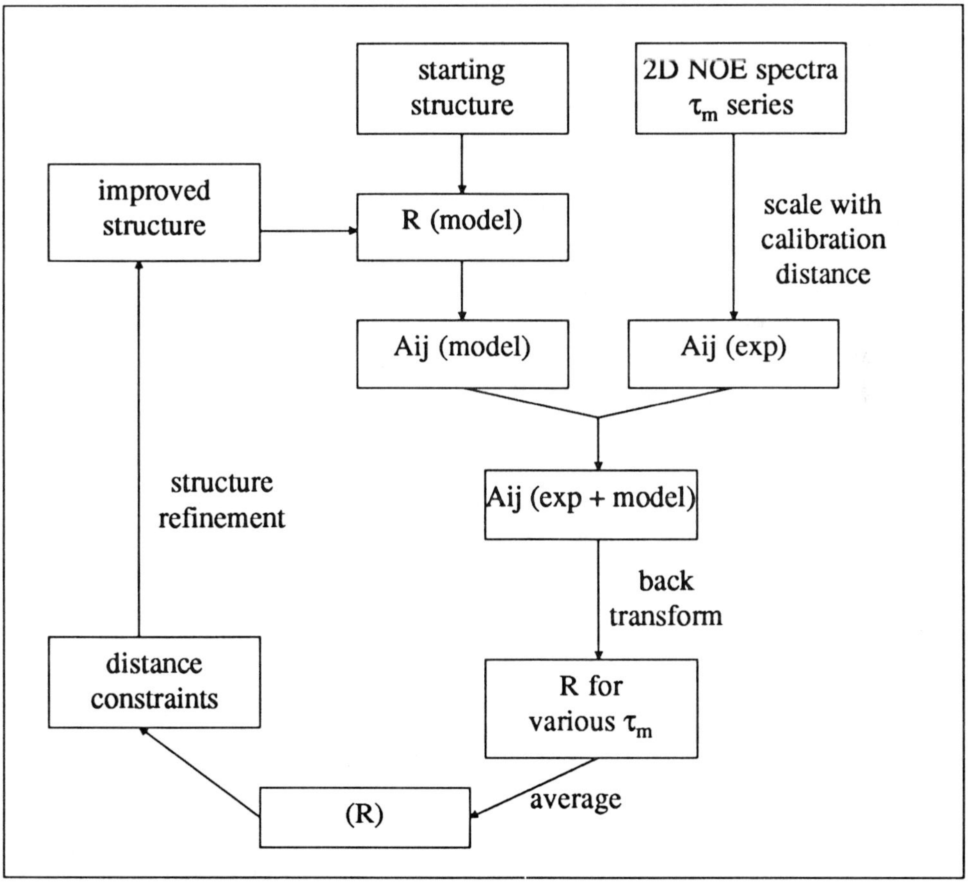

Fig. 2 Flow diagram for the iterative relaxation matrix approach (IRMA) for structure determination of biomolecules from proton-proton NOE's. *Reprinted from Ref. 9.*

series of mixing times this then leads to an improved set of distance constraints, from which a better structural model can be obtained using distance geometry or restrained molecular dynamics methods. The procedure can be repeated until all experimental NOE's are satisfactorily explained.

We have tested the method with calculations on a model structure of a DNA octamer d(GCGTTGCG) · d(CGCAACGC) [9] and also on an experimental 2D NOE data set of the same oligonucleotide [10]. In the model calculations we compared the distances, which could be obtained from initial rate analysis, with the actual distances from the model and simulated various experimental conditions. It was shown that back transformation of a combined NOE matrix led to a more accurate estimate of the cross-relaxation rates for proton pairs, between which NOE's are observed, than could be obtained from initial rate analysis.

Analysis of the experimental data set yielded 167 cross-peaks, for which the build-up of NOE intensity was measured. The cross-peaks between cytosine H5 and H6 protons, which have a fixed distance of 2.45 Å, were used for scaling the experimental and calculated NOE matrices. In order to test the starting model dependence of the procedure both A-DNA and B-DNA starting conformations were used. For both models the NOE matrices were calculated by diagonalizing the 182×182 relaxation matrices. Then, 167 off-diagonal elements were replaced by the measured ones. Note that the strong diagonal intensities are taken from the model calculation. These combined experimental/model NOE matrices are then back transformed to give new relaxation matrices and (assuming isotropic rigid motion) a new set of proton proton distances now corrected for spin-diffusion. The next step was a restrained MD refinement using the new set of distance constraints to arrive at an improved structure.

We could show [10] that starting from a given model (A- or B-DNA) convergence was obtained within three cycles of the IRMA procedure. This convergence occurs both for the proton proton distances (a new correction for spin-diffusion is calculated each time) and at the level of the three-dimensional structures. Also, starting from A- and B-DNA the structures converged to each other, giving final structures that are in the B-family. Thus, comparing the rms differences of the atomic coordinates of the A-DNA derived structure with the initial energy minimized A-DNA structure a value of 6.6 Å was found, while the B-DNA derived structure was only 2.4 Å away from its starting conformation. The rms difference between the two final structures (after three cycles of IRMA) was 2.0 Å. Fig. 3 shows a stereo diagram of the final structure (from B-DNA). A striking result is the strong bend in the helix axis that occurs between basepairs 4 and 5. This was present in both A and B-DNA derived structures and may be an example of a so-called AA wedge [40]. All sugar rings have Cl'-exo or C2'-endo conformations as normally found for B-DNA.

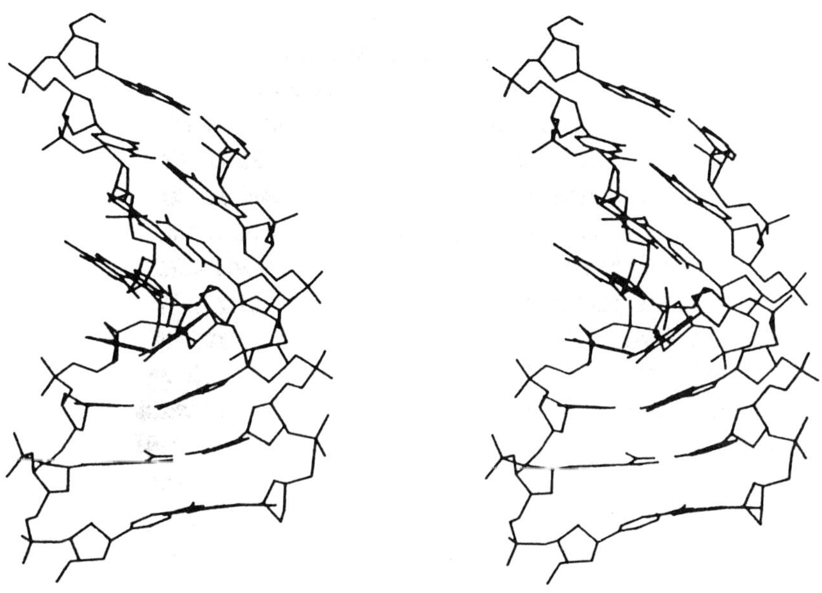

Fig. 3 Stereo pairs of the structure of the DNA octamer d(GCGTTGCG) · d(CGCAACGC) obtained after three cycles of IRMA starting from B-DNA.

Comparison of the experimental NOE build-up curves with those calculated on the basis of the converged structures showed that the cross-peaks of protons at short distances (2.0-3.0 Å) contain relatively little contributions of spin-diffusion. However, for the longer distances (3.5-4.5 Å) this contribution can be appreciable even at short (50 ms) mixing times. This could lead to errors in the distance of 20-25% in some cases. The relaxation matrix approach eliminates this kind of errors and yields a much more reliable set of distance constraints than can be obtained by the conventional initial rate analysis. It has the advantage that longer mixing times can be used, so that full advantage of the sensitivity of the spectrometer can be taken. Furthermore, longer proton proton distances can be determined even those for which the NOE cross-peaks are dominated by spin-diffusion.

5.4 Protein-DNA Interaction

5.4.1 *Lac* repressor headpiece structure

The first step in the structure determination of *lac* headpiece consisting of the 51 N-terminal amino acid residues of *lac* repressor was the sequential assignment of its ^1H resonances according to the procedure described by Wüthrich [2]. For all amide and C_α-protons except those of Ile 48 (due to an overlap problem) assignments could be made [41,42]. Most of the side chain protons have also been assigned, although for some of the long side chains of Lys, Arg and Gln the assignments do not extend beyond the C_β-protons [41,42]. By making use of a combination of J-coupling ($^3J_{\alpha\beta}$) and NOE's the pro-chiral methyl groups of valines 9, 20, 23 and 38 could be stereospecifically assigned [19].

Inspection of assigned cross-peaks in the 2D NOE spectra of *lac* headpiece showed that it is a typical α-helical protein with no β-sheets. In three regions of the protein relatively strong NOE's were found that correspond to short distances d_{NN}, $d_{\alpha N}(i,i+3)$ and $d_{\alpha N}(i,i+4)$ prevailing in α-helices [43]. The three α-helices are found in the regions 6-13, 17-25 and 34-45 of the polypeptide. Consistent with this is the observation of slow H-D exchange for the amide protons in the helical hydrogen bonds [44]. It should be noted that with the combination of NOE and exchange data the secondary structure of proteins can be established quite reliably.

The next step was the tertiary structure determination. This was done on the basis of 169 NOE's observed in 2D NOE spectra taken at a relatively short mixing time of 50 ms. As a distance calibration we used the NOE intensities of a series of cross-peaks corresponding to the distances $d_{\alpha N}$ and $d_{\alpha N}(i,i+3)$ of α-helical regions, which are both approximately 3.5 Å. Thus, the NOE's were essentially converted to distance ranges of 1.8-3.5 Å (the lower limit being the sum of the van der Waals radii). To these ranges pseudo-atom corrections were applied for groups that show dynamic averaging effects (methyl groups and tyrosine rings) and in cases where the stereospecific assignments of diastereotopically related protons were not known. To this data set 17 hydrogen bond constraints were added for those slowly exchanging amide protons for which the H-bond acceptor was known, i.e. for the α-helical regions. The initial structure determination was carried out with the restrained molecular dynamics method, using a model built conformation as the starting structure [7,11]. A relatively low value for the force constant K_{dc} of 250 kJmole^{-1}nm^{-2} was used, allowing for excess distances (d_{ij}–d_{ij}^0) of ca. 1 Å before an energy penalty is felt at the level of kT. However, larger values for K_{dc} up to 4,000 kJmole^{-1}nm^{-2} were sometimes found to speed up convergence. At a later stage the absence of NOE's between assigned protons was also taken into account [22]. Thus, repulsive pseudo-potential terms for a carefully selected set of 9529 non-NOE's were included. Although the number of non-NOE's may seem large, the information content is in fact rather low, since most non-NOE's are trivial in a structure that is already approximately correct. After a RMD run of 60 ps (without solvent) the resulting structure satisfied the experimental constraints very well and at the same time had a low energy (the energy dropped from +4074 kJmole^{-1} for the starting structure to –3091 kJmole^{-1} after RMD followed by REM). The remaining violations of the constraints were not larger than 0.5 Å, while the sum of all violations was 5.8 Å [22]. Fig. 4 shows a stereo picture of a snapshot taken from the RMD run. The

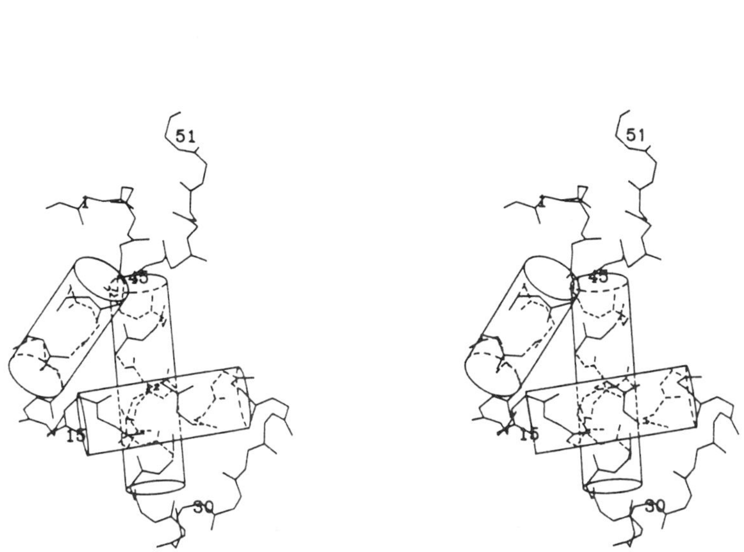

Fig. 4 Stereo diagram of the backbone conformation of headpiece 51. Cylinders represent the three α-helices. The structure was determined from a set of 169 proton proton distance constraints from NOE's using a restrained molecular dynamics procedure [7]. *Reprinted from Ref. 54.*

helix-turn-helix region consisting of the helices I and II of the headpiece can be clearly seen with the third helix packing against these forming a hydrophobic core. The RMD run also indicated that the three-helical core of the protein is rather rigid, whereas the N-terminal and C-terminal region and also the loop between helix II and III showed higher mobility.

Next, the question of uniqueness was addressed or, in other words, what is the range of conformations consistent with the constraints. A series of DG calculations was performed using the same set of distance constraints [29]. An overlay of 10 DG structures is shown in Fig. 5A. The variation among these structures can be expressed as an average rms difference of the C_α-atom coordinates, which was 1.4 Å for all C_α's and 1.1 Å for those of residues 4-47. It can be noticed in Fig. 5A that the N-terminal and C-terminal peptide regions have a preference for extended backbone conformations in spite of the fact that there are no long range NOE's in these regions that would fix the conformation with respect to the helical core. This is clearly an artefact of the metric matrix DG procedure. Starting with the DG structures DDD runs were carried out with 1,000 integration steps each (formally corresponding to 2 ps). Fig. 5B clearly shows that the N-terminal and C-terminal peptide fragment adopt a much wider range of conformations, while the structures satisfied the constraints equally well. The average rms differences of the DDD structures were 3.0 Å for all C-atoms and 2.0 Å for the C_α's of residues 4-47. Thus, the DDD procedure in combination with DG greatly improves the sampling properties compared to DG alone.

Finally, RMD refinement of these structures resulted in the family of structures shown in Fig. 5C. Some convergence can be noticed in helical regions, the rms difference for the C_α-atoms of residues 4-47 now becomes 1.7 Å (for the C_α's of the helices this value is 0.8 Å showing that the helical core of the protein is particularly well determined). Calculated for all C_α-atoms the rms difference is still 3.0 Å which is mainly caused by the large spread in conformations of the N-terminal and C-terminal peptides.

5.4.2 *Lac* Headpiece-operator Complexes

The *lac* headpiece structure formed the basis for further studies aimed at determining the way in which *lac* repressor recognizes its operator. *Lac* operator of *E.coli* is defined genetically as the control region in the *lac* operon, where

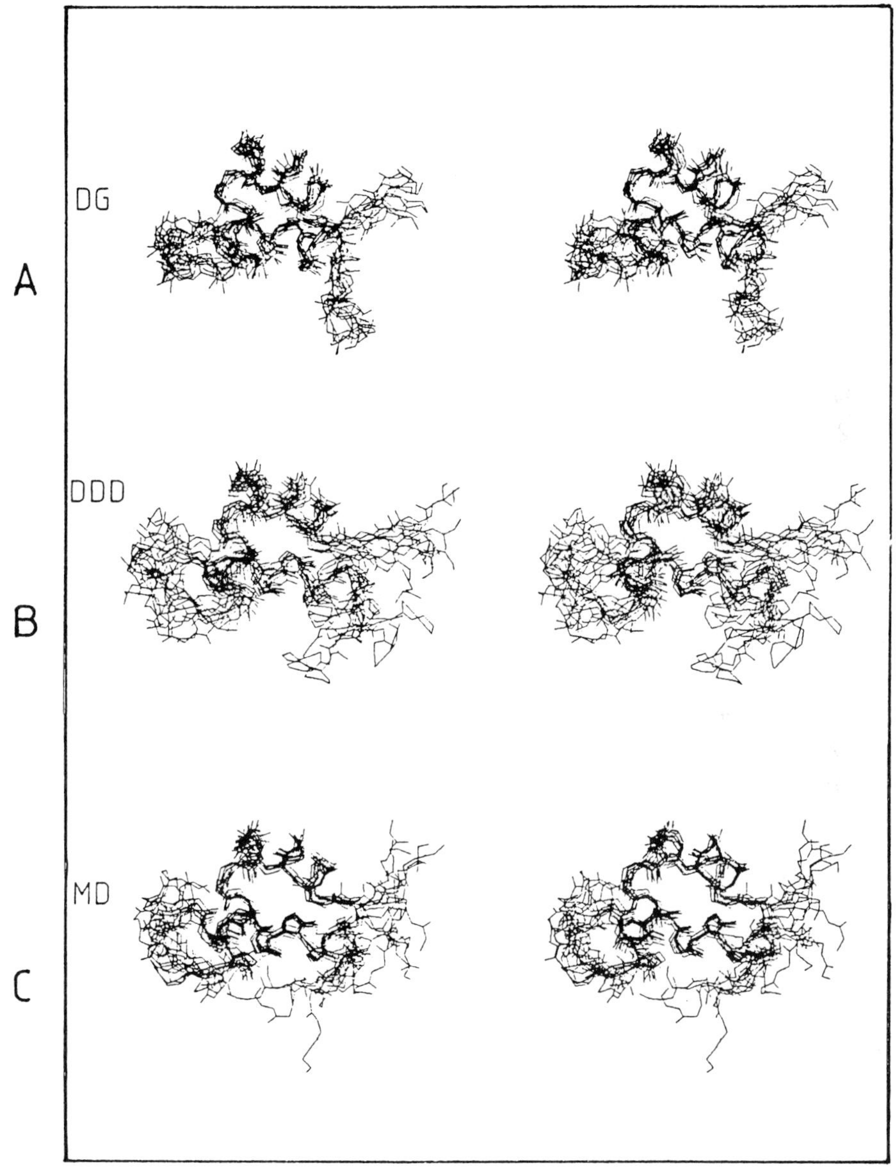

Fig. 5 Family of 10 conformations of *lac* repressor headpiece obtained after distance geometry (A), distance bounds driven dynamics (B) and the restrained molecular dynamics refinement (C). Stereo diagrams of the backbone atoms are shown. *Reprinted from Ref. 55.*

operator constitutive mutants occur. The region protected by *lac* repressor is 20-25 bp long with a pseudo dyad axis going through GC11 [45].

	–1	1 5				10					15	20		
TG	G	AATTGT	G	A	G	C	G	G	A	T	A	ACAATT	T	CA
—	•	—	•	—	•	—	•	—	•	—	•	—	•	—
AC	C	TTAACA	C	T	C	G	C	C	T	A	T	TGTTAA	A	GT
14–FRAGMENT														

Based on the hypothesis that *lac* repressor binds to the operator with two headpieces, thereby preserving the pseudo C_2 symmetry in the repressor-operator complex we have chosen a 14 bp half operator fragment for studies with single headpieces. The total molecular weight of the headpiece-14 bp operator complex is ca. 14,000, approaching in complexity the limit of what can be presently studied by 2D NMR at the level of individual resonance assignments for large numbers of residues. The first 2D NMR spectra of the complex were already reported in 1983 [46]. However, the detailed analysis took several years [12,13] and is still not finished.

Fig. 6 shows one of the most readily accessible regions of a 2D NOE spectrum of the complex of HP56 with the 14 bp operator. It contains a window, where only intra-DNA cross-peaks occur (H6/H8-H1' and cytosine H5-H6). These cross-peaks provided a start for the assignment of the DNA resonances as is shown in Fig. 6 for one strand by the lines connecting intra and internucleotide cross-peaks. In this way assignments were obtained from all non-exchangeable protons of the DNA in the complex except for some of the H5' and H5" protons. The general pattern of intra-DNA NOE's is still that of a B-DNA type conformation. Also, most of the ^1H resonance positions show small shifts upon complex formation with a maximum of 0.2 ppm for the H8 proton of G5 and the H1' proton of G7. These results are also consistent with the idea that small adjustments of the DNA conformation occur. These conformational changes, however, cannot yet be specified, but are likely to involve bending or unwinding or a combination of both.

Similarly, a large number of ^1H assignments has been made for the protein part of the HP56-14 bp operator complex. For this a combination of 2D NOE spectra and homonuclear Hartmann-Hahn (HOHAHA) spectra was used. The protons of many internal residues such as Leu 6, Val 9, Leu 45 and Tyr 47 have characteristic chemical shifts (distinct from random coil), which change very little upon complex formation. Furthermore ca. 80% of the long range NOE's could be identified for headpiece in the complex. This shows that the basic three-helical structure of headpiece is conserved when it binds to the operator. Shifts only occur for residues in the DNA binding site, which may be due to the presence of the DNA

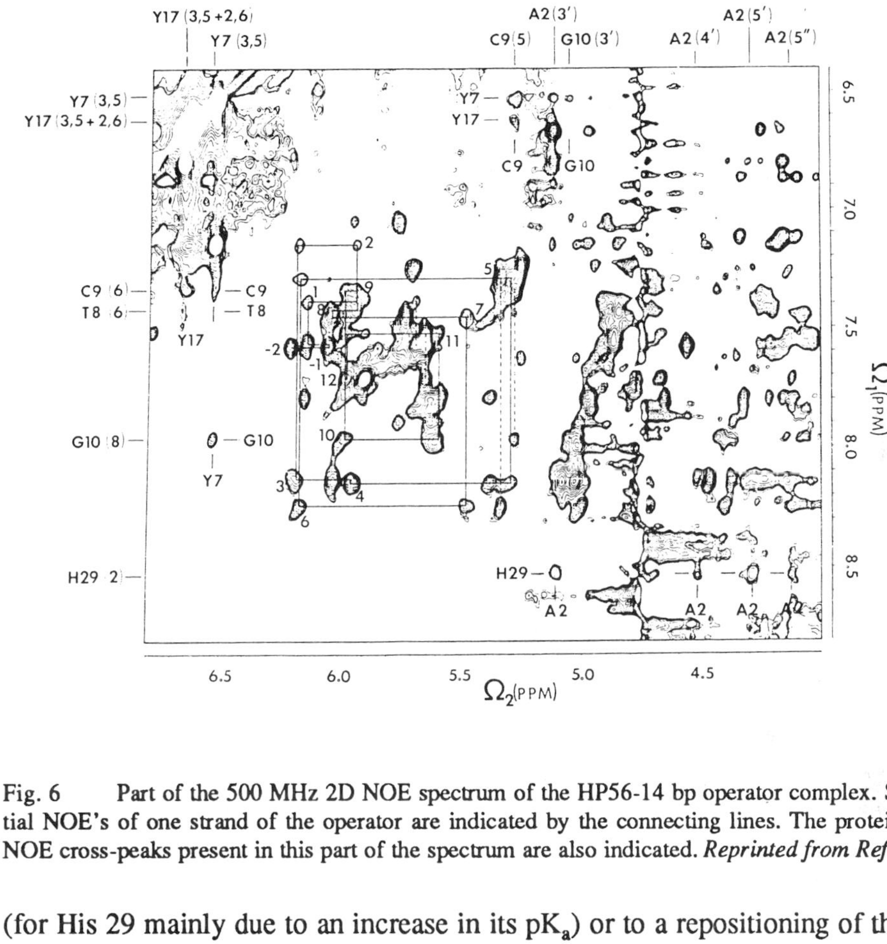

Fig. 6 Part of the 500 MHz 2D NOE spectrum of the HP56-14 bp operator complex. Sequential NOE's of one strand of the operator are indicated by the connecting lines. The protein-DNA NOE cross-peaks present in this part of the spectrum are also indicated. *Reprinted from Ref. 54.*

(for His 29 mainly due to an increase in its pK_a) or to a repositioning of the side chains as probably occurs for Tyr 7 and Tyr 17.

Using the assignments of the 14 bp operator and of HP56 in the complex it was possible to detect NOE's between protein and DNA. Some of these can be seen as cross-peaks in the 2D NOE spectrum of Fig. 6. For instance, extending the horizontal line at 7.96 ppm (of the H8 proton of G10) to low field one finds a cross-peak at 6.53 ppm, which can only belong to the 3,5 protons of Tyr 7. The analysis so far has yielded 24 inter protein DNA NOE's, which are listed in Table 1.

Tab. 1 NOE's between *lac* repressor headpiece 56 and a 14 bp *lac* operator fragment.

Protein	DNA	Protein	DNA
unambiguous [a]		probable [a]	
Tyr7 H3,5	G10 H8	Thr5 Cγ H3	G10 H8
Tyr7 H3,5	G10 H1'	Thr5 Cγ H3	G10 H3'
Tyr7 H3,5	G10 H3'	Leu6 Cδ H3	C9 H5
Tyr7 H3,5	C9 H5	Leu6 Cδ H3	C9 H6
Tyr7 H3,5	C9 H6	Leu6 Cδ H3	C9 H3'
Leu6 Cδ H3	C9 H5	Leu6 Cδ H3	T8 H6
Tyr17 H3,5 + H2,6	C9 H5	Tyr17 H3,5 + H2,6	T8 CH3
Tyr17 H3,5 + H2,6	C9 H6	Ser21 Cα H	T8 CH3
Tyr17 H3,5 + H2,6	T8 H6	Ser21 Cβ H	T8 CH3
His29 H2	A2 H8	His29 H2	A2 H3'
His29 H2	T3 CH3	His29 H2	A2 H4'
		His29 H2	A2 H5' [b]
		His29 H2	A2 H5" [b]

[a] The unambiguous NOE's were assigned at unique resonance frequencies, while the probable NOE's were from resonances which could overlap with resonances of other protons (see text for further discussion).

[b] H5' and H5" protons were only pair-wise assigned.

A distinction is made between NOE's that are unambiguous because they involve protons with unique resonance positions and those that are probable. The latter ones occur in crowded regions where overlap of resonances may occur. They were assigned on the basis of a pattern recognition procedure, which involves the following reasoning. Suppose a cross section of a headpiece proton shows NOE's to a set of other protons of the same amino acid residue. Then if a cross section of a DNA proton shows cross-peaks at the same frequencies and at least one of these can be uniquely assigned to a headpiece proton, we consider the assignment of the other cross-peaks in the set also extremely likely. The case of His 29 may serve as an example. In Fig. 6 a horizontal line at the C2 proton frequency of His 29 (8.62 ppm) shows a set of four cross-peaks in the DNA ribose region which are in a crowded region of the spectrum. Now, a similar set of cross-peaks is observed at the line of H8 of adenine 2 and, moreover, a very weak cross-peak is observed at the crossing of this line and that of His 29, which have both unique resonance positions (not shown). Hence, the weak NOE between H8 of A2 and the His 29 C2 proton is listed in the upper part of Table 1 although it undoubtedly is the result of spin-diffusion. The other NOE's of His 29 with the ribose protons of A2 are stronger and represent shorter distances, but occur in a region of the spectrum with much overlap and are therefore indicated as probable.

5.4.3 A Structural Model for the Headpiece-Operator Complex

Since we have shown that neither headpiece nor operator undergoes large conformational changes upon binding, it seemed reasonable to attempt building a model of the complex based on the structures of the constituents and the NOE's of Table 1. It should be realized that at this stage only a low resolution model of the complex can be obtained because the 2D NOE spectra were run under conditions of limited spin-diffusion and the distance constraints derived from them could be up to ca. 6 Å. The DNA was kept in the standard B-DNA conformation. For headpiece the backbone conformation as found by Zuiderweg et al. [11] was also kept rigid, but some of the side chains in the DNA binding site were allowed to change their conformation. It was then possible to obtain a model that satisfied all NOE constraints [12,13]. Energy minimization of this model (J. De Vlieg and R. Kaptein, unpublished results) was carried out to ensure that it has reasonable non-bonded interactions. A schematic picture of this model is shown in Fig. 7.

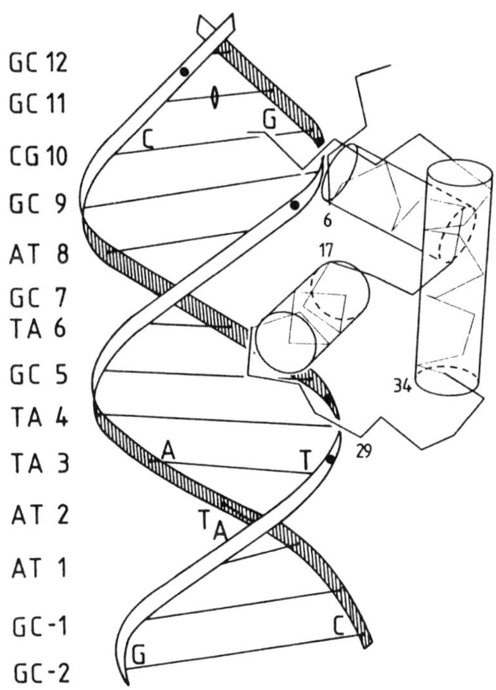

Fig. 7 Model of the headpiece-operator complex. The DNA is in the standard B conformation, while the backbone conformation of the headpiece is taken from Zuiderweg et al. [41]. The black dots indicate the phosphates where ethylation interferes with *lac* repressor binding [52]. ***Reprinted from Ref. 54.***

The fact that all NOE constraints are satisfied simultaneously is significant, because it means that the model must represent the major specific complex. Other, non-specific complexes that are undoubtedly formed at the high concentrations (5 mM) of the NMR experiments apparently do not lead to inconsistent NOE's either because they do not involve short proton proton distances or because each of them does not have a long enough life time to allow build up of NOE's. A 1:1 mixture of two different complexes would have led to inconsistent NOE's and can also be excluded.

The most surprising feature of the model is that the orientation of the second or "recognition" helix in the major groove of DNA with respect to the dyad axis at GC11 is opposite to what is found in all other models of repressor-operator interaction, either from direct X-ray observation as for 434 repressor [47] or from models built for CAP and λ and cro repressors [48]. Indeed, it is also opposite to orientations predicted for *lac* repressor on the basis of the analogy of models for CAP [49] and cro repressor [50]. In these models the first helix would be away from the dyad axis, while in the complex shown in Fig. 7 it is close to it. Our model accounts for the phosphate ethylation interference experiments (indicated in Fig. 7) and also for a functional contact between Gln 18 and GC 7 as found by Ebright [51] from a genetic "loss of contact" study involving mutants of both *lac* repressor and operator. We note, however, that all this and numerous other biochemical and genetic data (but not the NOE's) can be equally well explained by headpieces in the opposite orientation on the DNA.

The question, then, whether the whole *lac* repressor binds to operator with its headpieces in the same orientation as we have found for isolated headpiece is extremely interesting, but cannot be answered by NMR. If X-ray diffraction does not provide the answer, it has to come from genetic studies, where pairwise interactions between amino acid residues and DNA bases can be demonstrated. Some attempts in this direction are currently made by the group of Müller-Hill. Initial results were interpreted as weakly against the orientation of our HP-operator model [53], but more recent data seem to support it. Müller-Hill (personal communication) recently found a repressor mutant with Arg 22 replaced by Asn, which showed altered binding specificity towards a *lac* operator with GC5 replaced by TA. This result implies a Arg 22 - GC5 contact in the native system, that we predicted on the basis of the NMR results [13]. From the NMR work discussed here and from various model building studies on other repressors, the interesting conclusion emerges, that, although the helix-turn-helix motif seems to be universally used in DNA recognition by procaryotic repressors, it is used in different ways. It has been noted earlier [48] that the position of the recognition helix with respect to the major DNA groove varies much from parallel to the groove in the cro-O_R3 model to almost perpendicular in the CAP-DNA model. Now the *lac* headpiece-operator study shows that it can also be used in opposite orientations. Future work will be directed to specify more precisely the nature of the interactions between amino acid residues and base pairs. It is our hope that these and other NMR studies will contribute to a basic understanding of sequence specific recognition of DNA by proteins.

References

[1] Ernst, R.R., Bodenhausen, G. and Wokaun, A. "Principles of Nuclear Magnetic Resonance in One and Two Dimensions" Clarendon Press, Oxford, (1987).

[2] Wüthrich, K. "NMR of Proteins and Nucleic Acids" Wiley, New York, (1986).

[3] Noggle, J.H. and Schirmer, R.E. "The Nuclear Overhauser Effect-Chemical Applicatons" Academic Press, New York, (1971).

[4] Havel, T.F., Crippen, G.M. and Kuntz, I.D. *Biopolymers* (1979) *18*, 73-81

[5] Braun, W. and Go, N. *J. Mol. Biol.* (1985) *186*, 611-626.

[6] Van Gunsteren, W.F., Kaptein, R. and Zuiderweg, E.R.P. in "Nucleic Acid Conformation and Dynamics" Olson, W.K. Ed. Report of Nato/CECAM Workshop, Orsay, pp. 79-92 (1983).

[7] Kaptein, R., Zuiderweg, E.R.P., Scheek, R.M., Boelens, R. and van Gunsteren, W.F. *J. Mol. Biol.* (1985) *182*, 179-182.

[8] Clore, G.M., Gronenborn, A.M., Brünger, A.T. and Karplus, M. *J. Mol. Biol.* (1985) *186*, 435-455.

[9] Boelens, R., Koning, T.M.G. and Kaptein, R. *J. Mol. Struct.* (1988a) *173*, 299-311

[10] Boelens, R., Koning, T.M.G. van der Marel, G.A., van Boom, J.H. and Kaptein, R. *J. Magn. Res.* in the press (1988b).

[11] Zuiderweg, E.R.P., Scheek, R.M., Boelens, R., van Gunsteren, W.F. and Kaptein, R. *Biochimie* (1985c) *67*, 707-715.

[12] Boelens, R., Scheek, R.M., van Boom, J.H. and Kaptein, R. *J. Mol. Biol.* (1987a) *193*, 213-216.

[13] Boelens, R., Scheek, R.M., Lamerichs, R.M.J.N., de Vlieg, J., van Boom, J.H. and Kaptein, R. in "DNA-Ligand Interactions", Guschlbauer, W. and Saenger, W. Eds. Plenum, New York, pp. 191-215 (1987b).

[14] Braunschweiler, L. and Ernst, R.R. *J. Magn. Reson.* (1983) *53*, 521-528.

[15] Davis, D.G. and Bax, A. *J. Am. Chem. Soc.* (1985) *107*, 2821-2822.

[16] Jeener, J., Meier, B.H., Bachmann, P. and Ernst, R.R. *J. Chem. Phys.* (1979) *71*, 4546-4553.

[17] Bothner-By, A.A., Stephens, R.L., Lee, J., Warren, C.D. and Jeanloz, R.W. *J. Am. Chem. Soc.* (1984) *106*, 811-813.

[18] Bax, A. and Davis, D.G. *J. Magn. Res.* (1985) *63*, 207-213

[19] Zuiderweg, E.R.P., Boelens, R. and Kaptein, R. *Biopolymers* (1985a) *24*, 601-611.

[20] Hyberts, S.G., Marki, W. and Wagner, G. *Eur. J. Biochem.* (1987) *164*, 625-635.

[21] Keepers, J.W. and James, T.L. *J. Magn. Reson.* (1984) *57*, 404-426.

[22] De Vlieg, J., Boelens, R., Scheek, R.M., Kaptein, R. and van Gunsteren, W.F. *Isr. J. Chem.* (1986) *27*, 181-188.

[23] Neuhaus, D., Wagner, G., Vasak, M., Kägi, J.H.R. and Wüthrich, K. *Eur. J. Biochem.* (1985) *151*, 257-273.

[24] Blumenthal, L.M. "Theory and Applications of Distance Geometry", Chelsea, New York, (1970).

[25] Havel, T.F., Kuntz, I.D. and Crippen, G.M. *Bull. Math. Biol.* (1983) *45*, 665.

[26] Havel, R.F. and Wüthrich, K. *J. Mol. Biol.* (1985) *182*, 281-294.

[27] Havel, T.F. and Wüthrich, K. *Bull. Math. Biol.* (1984) *45*, 673-698.

[28] Wagner, G., Braun, W., Havel, T.F., Schaumann, T., Go, N. and Wüthrich, K. *J. Mol. Biol.* (1987) *196*, 611-639.

[29] De Vlieg, J., Scheek, R.M., van Gunsteren, W.F., Berendsen, H.J.C., Kaptein, R. and Thomason, J. *Proteins* (1988) *3*, 209-218.

[30] Scheek, R.M. and Kaptein, R. in "NMR in Enzymology", Oppenheimer, N.J. and James, T.L. Eds. Academic Press, New York, in press (1988).

[31] Brunger, A.T., Clore, G.M., Gronenborn, A.M. and Karplus, M. *Proc. Natl. Acad. Sci. USA* (1986) *83*, 3801-3805.

[32] Solomon, I. *Phys. Rev.* (1955) *99*, 559.

[33] Lefevre, J.F., Lane, A.N. and Jardetzky, O. *Biochemistry* (1987) *26*, 5076.

[34] Marion, D., Genest, M. and Ptak, M. *Biophysical Chemistry* (1987) *28*, 235.

[35] Broido, M.S., James, T.L., Zon, G. and Keepers, J.W. *Eur. J. Biochem.* (1985) *150*, 117

[36] Zhou, N., Bianucci, A.M., Pattabiram, N. and James, T.L. *Biochemistry* (1987) *26*, 7905.

[37] Brewer, J., Mendz, G.L. and Moore, W.J. *J. Am. Chem. Soc.* (1984) *106*, 4691.

[38] Perrin, C.L. and Gipe, R.K. *J. Am. Chem. Soc.* (1984) *106*, 4036.

[39] Olejniczak, E.T., Gaupe, R.T.Jr. and Fesik, S.W. *J. Magn. Reson.* (1986) *67*, 28.

[40] Trifonov, E.N. and Sussman, J.L. *Proc. Natl. Acad. Sci. USA* (1980) *77*, 3816-3820.

[41] Zuiderweg, E.R.P., Kaptein, R. and Wüthrich, K. *Eur. J. Biochem.* (1983a) *137*, 279-292.

[42] Zuiderweg, E.R.P., Scheek, R.M. and Kaptein, R. *Biopolymers* (1985b) *24*, 2257-2277.

[43] Zuiderweg, E.R.P., Kaptein, R. and Wüthrich, K. *Proc. Natl. Acad. Sci. USA* (1983b) *80*, 5837-5841.

[44] Boelens, R., Scheek, R.M., Dijkstra, K and Kaptein, R. *J. Mag. Res.* (1985) *62*, 378.

[45] Gilbert, W. and Maxam, A. *Proc. Natl. Acad. Sci. USA*, (1973) *70*, 3581-3584.

[46] Kaptein, R., Scheek, R.M., Zuiderweg, E.R.P., Boelens, R., Klappe, K.J.M., van Boom, J.H., Rüterjans, H. and Beyreuther, K. In: Clementi, E., Sarma, R.H. Eds. "Structure and Dynamics: Nucleic Acids and Proteins". Adenine Press, New York, pp. 209-225 (1983).

[47] Anderson, J.E., Ptashne, M.G. and Harrison, S.C. *Nature* (1987) *326*, 846-849

[48] Pabo, C. and Sauer, R. *Ann. Rev. Biochem.* (1984) *53*, 293-321.

[49] Weber, I.T., McKay, D.B.M. and Steitz, T.A. *Nucl. Acids. Res.* (1982) *10*, 5085-5102.

[50] Matthews, B.W., Ohlendorf, D.H., Anderson, W.F. and Takeda, Y. *Proc. Natl. Acad. Sci. USA* (1982) *79*, 1428-1452.

[51] Ebright, R.H. *J. Biomolec. Struct. Dyn.* (1985) *3*, 281-297.

[52] Barkley, M.D. and Bourgeois, S. In "The Operon", Miller, J.H. and Reznikoff, W.S. (Eds) Cold Spring Harbor Press, pp. 177-220 (1978).

[53] Lehming, N., Sartorius, J., Niemöller, M., Genenger, G., van Wilcken-Bergmann, B. and Müller-Hill, B. *EMBO. J.* (1987) *6*, 3145-3153.

[54] Eckstein, F. and Lilley, D.M.J Eds "Nucleic Acids and Molecular Biology 2" Springer-Verlag, pp. 167-187 (1988).

[55] Kaptein, R., Boelens, R., Scheek, R.M. and van Gunsteren, W.F. *Biochemistry* (1988) *27*, 5389-5395.

6. ^{31}P and ^{1}H 2D NMR and NOESY-Distance Restrained Molecular Dynamics Methodologies for Defining Structure and Dynamics of Wild-Type and Mutant Lac Repressor Operators. Sequence-Specific Variations in Double Helical Nucleic Acids.

David G. Gorenstein, Claude R. Jones, Stephen A. Schroeder, James T. Metz, Josepha M. Fu, Vikram A. Roongta, Robert Powers, Christine Karslake, Edward Nikonowitz and Robert Santini.

6.1 Introduction

Transcriptional control of gene expression is fundamentally important in cellular growth function, and development. Understanding the detailed structure and dynamics of protein-nucleic acid interaction is central to understanding many fundamental problems in biology and medicine, including carcinogenesis and viral infection. The detailed structural mechanism by which proteins bind to specific sequences of DNA is an area of great current interest.

It is now widely appreciated that duplex DNA can exist in a number of different conformations [1]. Significant conformational differences can exist globally along the entire double helix, as in the A, B, C and Z forms of DNA. In addition, local conformational heterogeneity in the deoxyribose phosphate backbone has been most recently noted in the form of sequence-specific variations [2,3] or as the result of drug [1] or protein binding [4] to local regions of the DNA.

It has been suggested that these localized, sequence-specific conformational variations are quite likely an important component of a DNA binding protein's recognition of specific sites on the DNA [3,5]. Thus, although the *lac* repressor protein does not recognise an alternating AT sequence as part of the lac operator

DNA sequence, the repressor protein binds to poly d(AT) 1000 times more tightly than to random DNA [1]. The repressor protein is quite likely recognizing the alternating deoxyribose phosphate backbone geometry of the two strands [6], rather than the chemical identity of the AT base pairs. Indeed, a recent X-ray crystallographic study [4] on a related repressor-DNA complex shows that much of the specificity of the protein for the operator DNA sequence derives from protein contacts with the *phosphate backbone* and thus much of the operator-DNA specificity may well originate from sequence-specific variation in the conformation of the DNA.

Until recently, [7-9] NMR spectroscopy has not been particularly successful in defining these potentially significant sequence-specific variations in the local conformation of the DNA [10-13]. Indeed a recent ^{1}H NMR study of a duplex decamer failed to reproduce sequence-specific variations in the sugar ring conformation predicted by X-ray crystal analysis [13]. Molecular mechanics and dynamics theoretical calculations [14-16] have also failed to reproduce these variations, and have raised the question whether the sequence-specific structural variations observed in the X-ray crystallographic studies are the result of less profound crystal packing forces. In this paper we will describe some of our recent efforts in utilizing NMR and molecular mechanics/dynamics calculations to define sequence-specific structural variations of nucleic acids and nucleic acid complexes.

6.2 Structural Studies of Oligonucleotides by 2D ^{1}H NMR

While X-ray crystallography has provided much of our understanding of these DNA structural variations, increasingly, high resolution NMR has also begun to provide detailed three-dimensional structural information on duplex oligonucleotides. Two dimensional NMR methods, [17,18] have allowed the identification of the base ^{1}H and most of the sugar ^{1}H signals. We have been able to use this methodology to directly assign almost all of the deoxyribose protons of the self complementary 14-mers 3-8 by $^{1}H/^{1}H$ COSY and NOESY spectroscopy (Figs. 1 and 2).

dCGCGAATTCGCG dGCGCTTAAGCGC 1	dCGTGAATTCGCG dGCGCTTAAGTGC 2

dTGTGAGCGCTCACA
dACACTCGCGAGTGT
3
(lac operator pseudo fragment)

lac operator pseudo fragment mutants:

dTATGAGCGCTCATA dATACTCGCGAGTAT 4	dTCTGAGCGCTCAGA dAGACTCGCGAGTCT 5
dTGTGTGCGCACACA dACACGCGCGTGTGT 6	dTGTGAGCGCTCACA dACACTCGCGAGTGT 7

NMR Methodology. Assignments of the proton signals of oligonucleotide duplexes is accomplished through analysis of the two-dimensional COSY and NOESY NMR spectra following a sequential assignment methodology [19,20]. In the COSY spectrum (Fig. 1) of the 14 bp duplex, 3, d(TGTGAGCGCTCACA)$_2$, scalar couplings between protons are manifested as

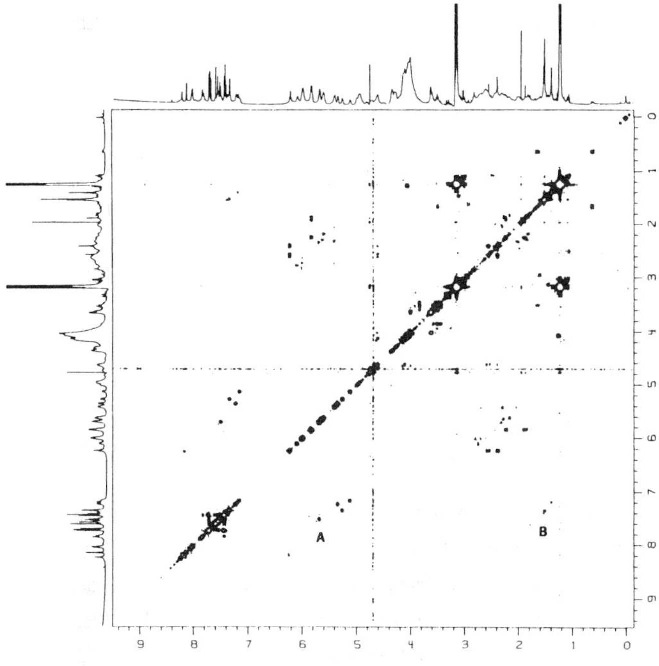

Fig. 1 Absolute value ^1H-^1H COSY NMR spectrum of duplex 14-mer, d(TGTGAGCGCTCACA)$_2$ at 470 MHz. *Reprinted from Ref. 8.*

off-diagonal cross-peaks. Each of the four cytosines, in one of the symmetrical halves of the spectrum, gives rise to a cross-peak representing the H5-H6 coupling. Similarly, the long range H6-CH_3 couplings in the three thymines give rise to cross-peaks. No cross- peaks from the purine bases are present since adenine and guanine do not possess groups of coupled non-exchangeable protons. The various couplings among the deoxyribose protons, i.e. H1'-H2', 2", H2'2"-H3', H3'-H4', H4'-H5', 5" may be traced through their COSY connectivities (although in longer oligonucleotides the cross-peaks may be very weak or even absent and connectivity must be established through the NOESY spectrum). While the COSY spectrum can be used to assign the protons on a particular base or sugar, it does not provide any information on the relative position of the base or sugar in the 14-mer sequence. This information can be obtained, however, through analysis of the cross-relaxation networks delineated in the NOESY spectrum (Fig. 2). Because the NOESY experiment [21] utilizes through-space connectivities rather than through-bond connectivities, correlations between base and sugar protons on neighbouring residues can be seen. This NOE information, taken together with the type of base assignments from the COSY spectrum, and com-

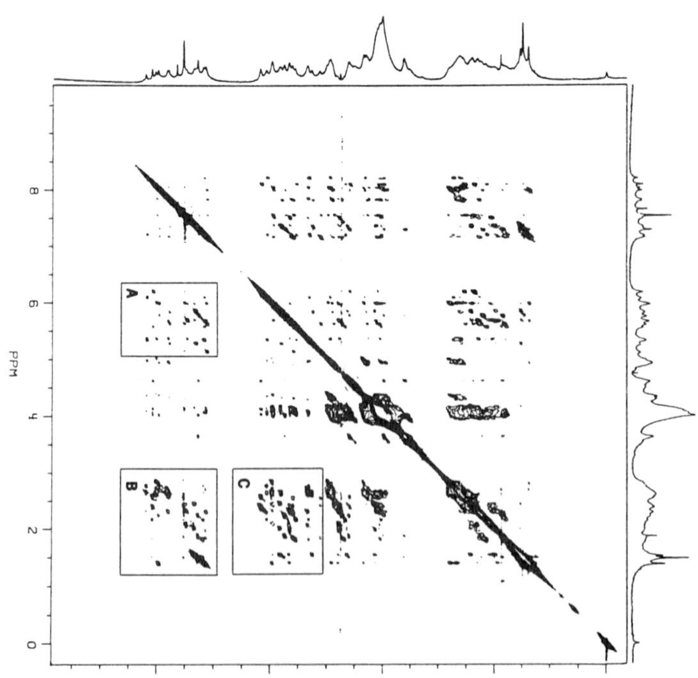

Fig. 2 Pure absorption phase ^1H-^1H NOESY NMR spectrum of duplex 14-mer, at 470 MHz. Region labelled A is expanded in Fig. 3. *Reprinted from Ref. 9.*

pared to the known sequence of the 14-mer, permitted assignment of nearly all the protons in the *lac* pseudo-operator segment [20]. In B-DNA, each pyrimidine H6 or purine H8 base proton is spatially situated so as to give rise to an NOE correlation with the H1' sugar proton of the same nucleotide as well as with the H1' sugar proton of the adjacent nucleotide on a 5' side (region A). The base at the 5' end of the chain can be identified by its lack of an NOE cross-peak to a 5' neighbouring sugar H1' proton and the sugar H1' proton at the 3' end can be identified by its lack of an NOE cross-peak to a 3' neighbouring base proton. Using these as starting points, it was possible to step through the entire helix via the sugar-base connectivities, as diagrammed in Fig. 3. The thus assigned sugar H1' protons are, in turn, correlated to the H2' and 2" protons of the same nucleotide, allowing these to be assigned [20]. Once the assignments of the sugar H1' and H2', 2" and the base protons were known, the H3' protons were identified through their NOE correlations to these protons. The H4' protons were assigned through their connectivities with H1'.

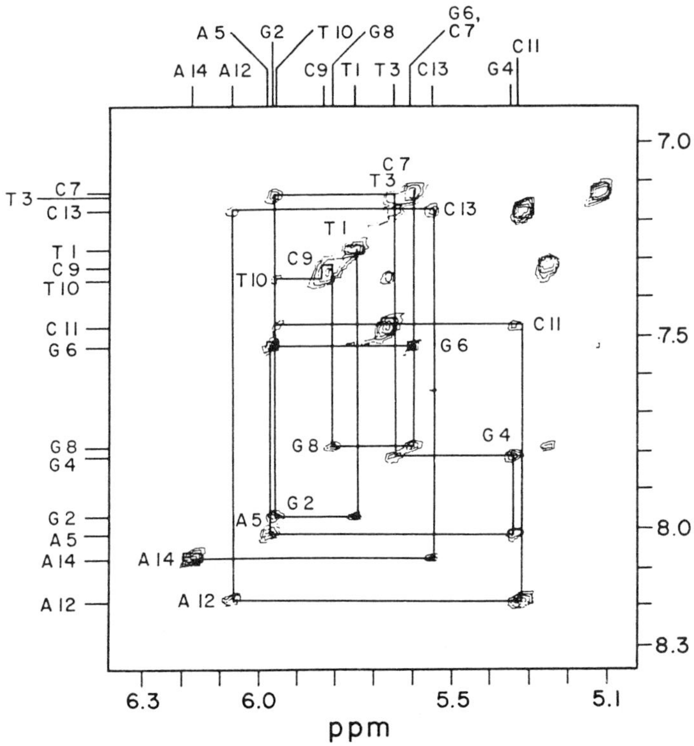

Fig. 3 Expansion of region A of the pure absorption phase ^1H-^1H NOESY spectrum shown in Fig. 2. The sequential assignment of the base and deoxyribose H1' protons is diagrammed. *Reprinted from Ref. 9.*

6.2.1 Restrained Molecular Mechanics Calculations of Duplex Geometries.

The relative intensities of the 2D NOESY cross-peaks, in the absence of spin diffusion and significant internal local motion, can provide information on the interatomic distances in macromolecules [22,23]. The theory for how we can obtain accurate distances is now well developed [24,25]. In brief, at very short mixing times, in the absence of spin diffusion, the rate of build-up of NOESY cross-peak intensities is proportional to the inverse sixth power of the distance between the cross-relaxing protons. Currently two methods are often used to derive interatomic distances from the NOESY data: 1) simplified 2-spin approximation using a single NOESY dataset at short mixing times* [25,26] and 2) a complete relaxation matrix approach (CORMA [24]). In addition, two main approaches are being used to solve biomolecular structures from NOESY-derived distances: 1) distance geometry [25,27,28] 2) interactive computer graphics combined with restrained molecular dynamics [22].

We have utilized the restrained molecular mechanics/dynamics method in order to derive structures for our oligonucleotides [29]. Using the distances derived from the 2D NOESY spectrum of Fig. 2 (distances were calculated by integrating the cross-peaks and utilizing the two-spin approximation at short mixing times), in conjunction with restrained molecular mechanics and dynamics calculations, we derived a structure for the duplex 14-mer.

Starting from an idealized model-built duplex structure, we have used the molecular mechanics/dynamics energy minimisation program AMBER [30] to minimize the energy of the 14-mer with and without the NOESY-derived distance restraints (Fig. 4).

Instead of a simple harmonic potential error function to restrain the NMR-derived distances, we have modified AMBER so as to provide a flat well harmonic function which better reflects the intrinsic accuracy of these NOESY distance restraints [23]. As noted in Fig. 4, while some differences are observed between the restrained and unconstrained minimized duplexes (particularly at the ends of the duplex where fraying of the strands is possible), the main features of the B-DNA duplex are restrained in both the dynamics and mechanics calculations.

As an illustration of the importance of NOESY-distance restraints in the calculation of duplex structures, we plot the calculated helix twist as a function of se-

---

* must first be established that during this mixing time, the rate of build-up of NOESY cross-peak intensity is approximately linear.

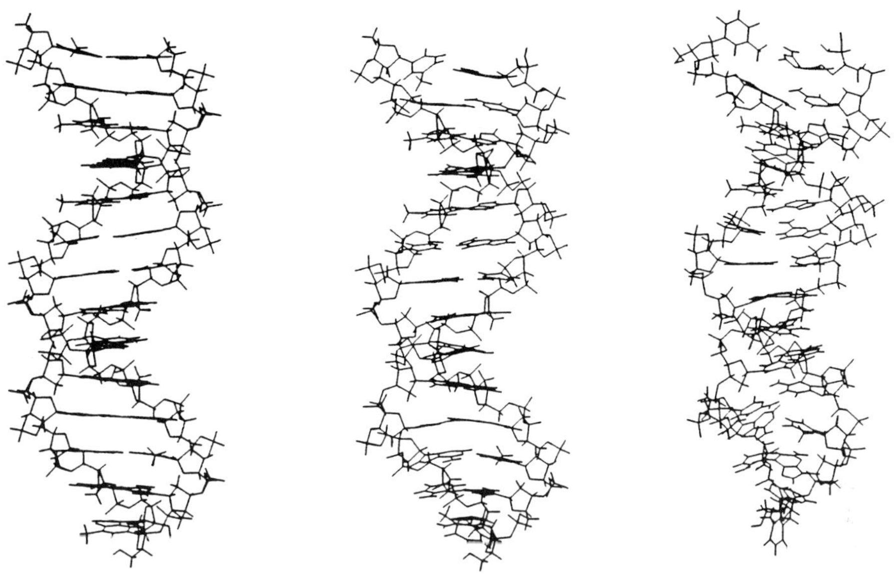

Fig. 4 (A) Molecular mechanics energy minimized structure of B-DNA model built 14-mer oligonucleotide using the AMBER program. (B) NOESY-distance restrained energy minimized structure. (C) NOESY-distance restrained molecular dynamics structure. *Reprinted from Ref. 8.*

quence for "wild-type" *lac* operator, 3 (Fig. 5). As described in more detail below, local helical distortions arise along the DNA chain due to purine-purine steric clash on opposite strands of the double helix [2]. As a result, 5'-PyPu-3' sequences within the oligonucleotide represent positions where the largest helical distortions occur. Dickerson [31,32] has shown that these sequence-specific variations in the conformation of duplex DNA observed in the crystal structure of GC-12-mer, 1, could be quantitatively predicted through a series of simple "Calladine rule" [2] sum function relationships. Thus the global helical twist (t_g) (defined in Fig. 6) can be calculated from eqn. 1 using the helical twist sum function (Σ_1):

$$t_g = 35.6° + 2.1\Sigma_1 \quad (1)$$

Dickerson [31,32] has shown that when correction is made for some crystal packing distortions and end-for-end averaging is used, the linear regression correlation coefficient between the calculated t_g from eqn. 1 and the crystallographically observed helical twist is > 0.9. Similar correlations have been established for base-plane roll angles, main-chain torsion angles δ, and propeller twist using Σ_2,

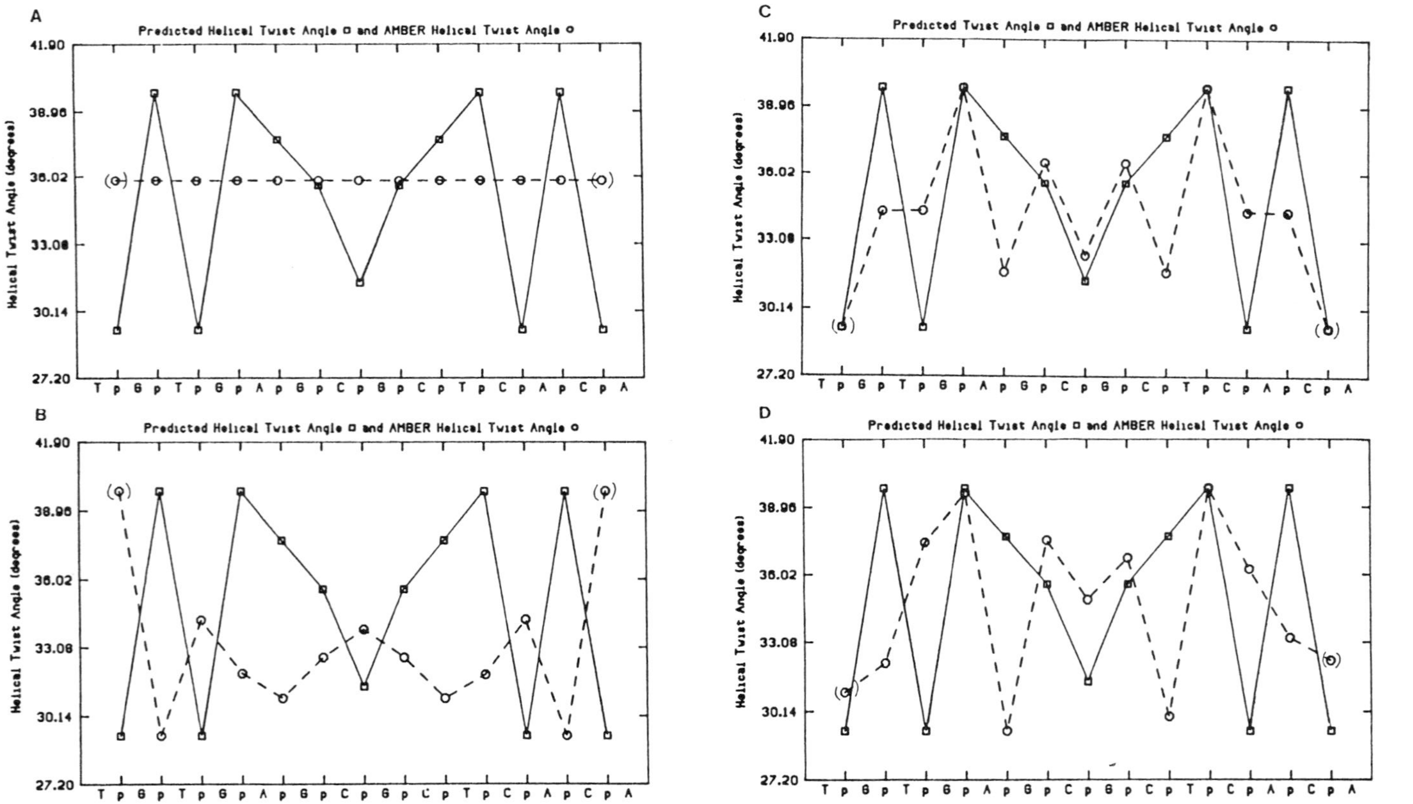

Fig. 5 Predicted sequence-specific variations in the helix twist values for "wild-type" *lac* operator, 3 (dashed curves calculated from helical twist sum function and eqn. 1). Helix twist values derived from calculated structures (solid curves): (A) AMBER molecular mechanics calculated structures and the helix twist values for these structures. (B) NOESY-distance restrained molecular mechanics calculated values for helix twist (C) NOESY-distance restrained molecular dynamics calculated values for helix twist as a function of sequences. *Reprinted from Ref. 9.*

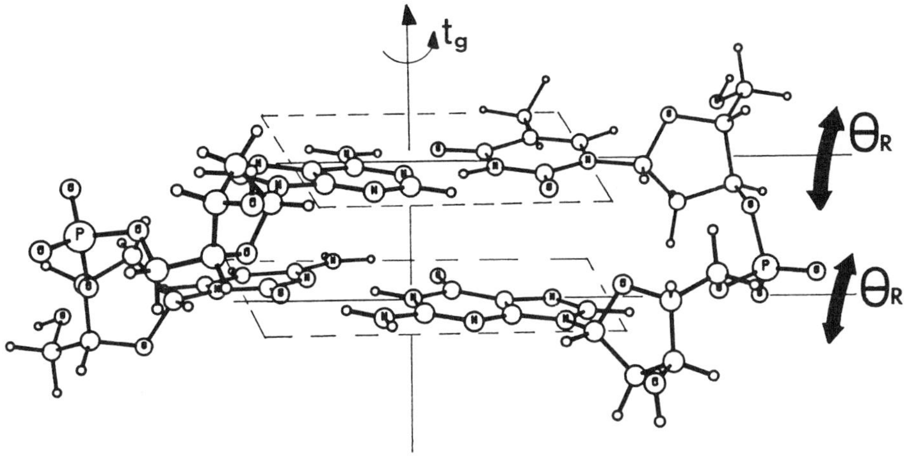

Fig. 6 Model of dTpG · dCpA showing definition of helix twist, t_g, and helix roll (Φ). Dickerson [30] defines t_g as the angle of the C1'-C1' vectors in two successive base pairs, viewing the duplex down along the best overall helix axis. In our own calculations we have followed the AMBER definition [30] based upon the N9-N9 vectors. Helix roll, Φ, is defined [31] as the change in the orientation of the best planes of successive base pairs along their long axes. *Reprinted from Ref. 9.*

Σ_3 and Σ_4 sum functions respectively. Molecular mechanics and dynamics theoretical calculations [14-16] have failed to reproduce these variations. This is illustrated in Fig. 5 where our molecular mechanics energy minimisation of a uniform B-DNA duplex geometry for the 14-mers fails to reproduce these sequence-specific variations. We start from a model built B-DNA geometry (based upon the fiber diffraction coordinates of Arnott and Hukins [33]), with uniform 35.8° t_g per base step (Fig. 5A; dashed curved). Also shown in Fig. 5A-D is the predicted t_g based upon the Dickerson/Calladine Σ_1 sum function relationship and eqn. 1 (solid curve). These rules predict that the helix twist should vary between 28° and 40° for this sequence, substantially different from the uniform B-DNA value of 35.8°. Energy minimization alone, using the AMBER program, does produce some variations in the helix twist, t_g for this sequence (Fig. 5B, dashed curve). Note however, that AMBER predicts a variation in t_g that is almost directly *opposite to* that predicted by the Calladine rules (correlation coefficient *negative* 0.86). Presumably this failure reflects the inability of an energy minimization program to locate the global energy minimum which would show the correct sequence-specific helix variations. This is a common problem with energy minimisation schemes and is attributable to the structure being trapped in a local energy minimum [30]. Of course, an alternative explanation for the discrepancy between the AMBER predicted values for helix twist and the Calladine rules, is that the Calladine rules are inapplicable.

We can "jump" out of these local energy traps by incorporating distances derived from our NOESY spectra. Using a flatwell pseudo-potential energy error function (eqn. 2) to describe the measured distances between various protons in our 14-mers, and incorporating this pseudo-energy penalty function into the AMBER molecular mechanics/dynamics program, we were able to calculate the structure for 14-mer, 3 shown in Fig. 4B. The skewed, flatwell harmonic potential E_{NOESY} has the following form:

$$E_{NOESY} = \begin{cases} k_L(r - r_0)^2 & r < (r_0 - L_{err}) \\ 0 & (r_0 - L_{err}) \leq r \leq (r_0 + R_{err}) \\ k_R(r - r_0)^2 & r > (r_0 + R_{err}) \end{cases} \qquad (2)$$

The left and right force constants are k_L and k_R, respectively. We have generally used an estimated error of ± 15% in the r_0 NOESY distances. The calculated helix twist values for this NOESY-distance restrained, energy minimized structure is shown in Fig. 5C. There is some improvement in the ability to reproduce the Dickerson/Calladine rule predicted values for the sequence-specific variations in the helix twist, although the correlation between calculated and predicted values are still quite poor (correlation coefficient between the Calladine rules and AMBER calculated helix twist values is now *positive* 0.335). No improvement in the fit between calculated and predicted helix twist values are obtained by simulating the structure in a NOESY-distance constrained molecular dynamics calculation (see Figs. 4C and 5D). Molecular dynamics calculations should be better able to overcome small energy barriers (on the order of kT) that otherwise limit the ability of an energy minimization scheme to locate a global energy minimum. By using "high" simulation temperatures (rt) and large "force constants" k_L and k_R (10 Kcal/mol/Å^2) for the NOESY-distance restraints, we are able to search for structures that better represent the "correct" solution conformation. The poorer fit of the data with the restrained molecular dynamics calculation relative to the restrained molecular mechanics calculation could be attributable to the lack of averaging of structures in the MD simulations (the MD structure shown above is a "snap shot" of the molecule at 0.3 ps). It has been shown [7,8] that NOESY distances appear to be able to restrain the calculated structures to conformations that accurately reflect these sequence-specific variations in the local conformation of the DNA. Our own data suggest that the NMR measured distances do not entirely reproduce the predicted variations in duplex geometry. Another ^{1}H NMR study of a duplex decamer also failed to reproduce sequence-specific variations in the sugar ring conformation predicted by X-ray crystal analysis [13]. The previous failure of molecular mechanics and dynamics theoretical calculations [14-16] to reproduce these variations and our own mixed results (including those described below utilizing ^{31}P NMR chemical shifts and coupling constants) raises the question whether the sequence-specific structural variations observed in the X-ray crystallographic studies are at least partially attributable to less profound crystal packing force. Thus, it is not entirely clear that these sequence-specific variations represent an intrinsic property of the nucleic acid structure in solution.

6.3 ^{31}P NMR of Nucleic Acids; Sequence-Specific Variations in Structure

While ^{1}H NMR can provide detailed information on the overall conformation of the deoxyribose rings and bases of oligonucleotides, it generally is unable to provide very much information on the phosphate ester backbone conformation. Of the six torsional angles that largely define the backbone structure, only the four involving the deoxyribose ring are amenable to analysis by ^{1}H NMR techniques (via coupling constant or NOESY-distance measurements). It has been suggested that the deoxyribose ring and base form a rather rigid unit with the main conformational flexibility of the nucleic acid backbone being limited to the two P-O phosphate ester torsional angles [34]. Thus, one of these, the C3'-O3'-P-O5' torsional angle, ζ, is found to be the most variable one in the B-form of the double helix and the other O3'-P-O5'-C5', α, torsional angle is one of the most variable in the A-form of the duplex [1]. Indeed following the original suggestion of Sundaralingam [34], and based upon recent X-ray crystallographic studies of oligonucleotides, Saenger [1] has noted that the P-O bonds may be considered the "major pivots affecting polynucleotide structure".

We have proposed that ^{31}P NMR spectroscopy is potentially capable of providing information on the most important remaining two torsional angles involving the phosphate ester bonds that define the nucleic acid backbone. Our studies [35-38] indicated that a phosphate diester monoanion in a gauche, gauche (*g, g*) conformation should have a ^{31}P chemical shift 1.5-2.5 ppm upfield from an ester in a non-*g, g* conformation. Our earlier ^{31}P NMR studies on poly- and oligonucleic acids [35,36,39-48] confirmed the suggestion that the base stacked, helical structure with a gauche, gauche phosphate ester torsional conformation should be upfield from the random coil conformation.

6.3.1 ^{17}O-Labelling methodology for Assigning ^{31}P Signals of Oligonucleotides

We have recently developed a phosphoramidite ^{17}O-labelling methodology to assign the ^{31}P signals of the phosphates in a number of the duplexes 2-7 [20,29,35,50-52]. Using a manual solid-phase phosphoramidite method [53] we can readily introduce ^{17}O or/^{18}O-labels in the phosphoryl groups by replacing the I_2/H_2O in the oxidation step of the phosphite by $I_2/H_2{}^{17}O$ (40%) or $H_2{}^{18}O$. By synthesizing a mono-^{17}O phosphoryl labelled oligonucleotide (one specific phosphate is labelled along the strand), we can identify the ^{31}P signal of that phosphate diester.

This is because the quadrupolar ^{17}O nucleus (generally ca. 40% enriched) broadens the ^{31}P signal of the directly attached phosphorus to such an extent that only the high resolution signal of the remaining 60% non-quadrupolar broadened phosphate at the labelled site is observed [36,54,55]. In this way each synthesized oligonucleotide with a different monosubstituted ^{17}O-phosphoryl group allows identification of one specific ^{31}P signal and the full series of monolabelled oligonucleotides gives the assignment of the entire ^{31}P NMR spectrum.

Figs. 7 and 8 show representative ^{31}P spectra of unlabelled and singly ^{17}O labelled GT-12-mers, 2 and 14-mer, 3. Fig. 7B is a spectrum of the GT-12-mer, 2, without any ^{17}O label that can be used as a comparison between labelled spectra.

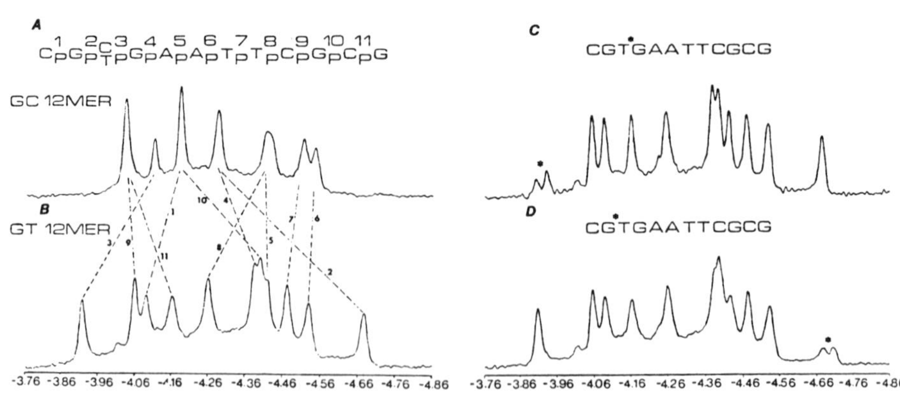

Fig. 7 ^{31}P NMR spectra and phosphate assignments of (A) GC-12-mer 1 [49] and (B) GT-12-mer 2. (Numbering corresponds to phosphate position from the 5'-end of the duplexes). Examples of site-specific ^{17}O labelling of two of the phosphates of GT-12-mer 5 at position 3 and 2 (shown by an asterisk) are shown in (C) and (D) respectively. Note the reduction in intensity of the ^{17}O labelled peaks. *Reprinted from Ref. 9.*

As can be seen in Figs. 7C/D, a decrease in intensity of a single resonance is observed. All eleven resonances can clearly be distinguished, each integrating for one phosphorus resonance. It is interesting to note that the resonance of the labelled phosphate is observed as two reduced intensity, resolved peaks associated with ^{16}O (unlabelled) and ^{18}O-labelled phosphorus resonances (The $H_2{}^{17}O$ sample also contains both $H_2{}^{16}O$ and $H_2{}^{18}O$).

This can most easily be seen in Figs. 7C and D where the ^{18}O-labelled phosphate ^{31}P signal is shifted slightly upfield relative to the remaining ^{16}O phosphorus resonance [36].

Patel and coworkers [56] have shown that base-pair mismatch in duplex 2 provides some very interesting ^{31}P spectral shifts. Whereas the ^{31}P spectral dispersion is < 0.7 ppm in normal B-DNA double helices (as seen in the spectra of Figs. 7 and 8 for duplex oligonucleotides 1 and 3), new signals are shifted upfield

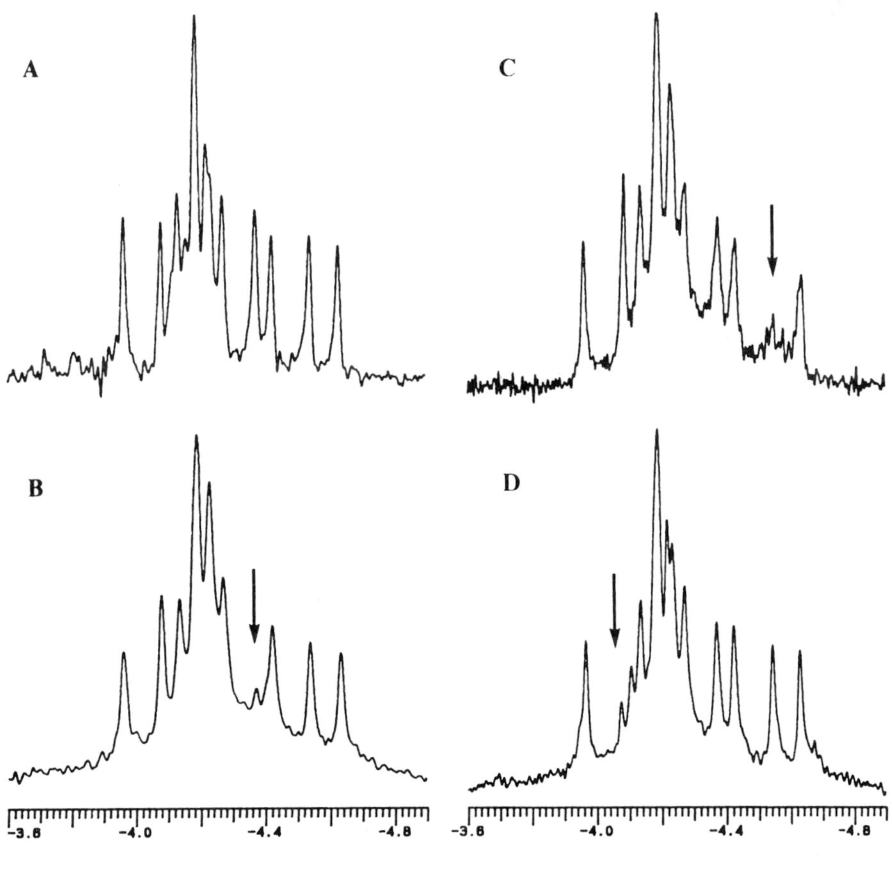

Fig. 8 ^{31}P NMR spectra of 14-mer 4 (A) and ^{17}O labelled 14-mer at indicated position. Note the reduction in intensity of the ^{17}O labelled peaks. *Reprinted from Ref. 9.*

and downfield from the "normal" double helical phosphate ^{31}P signals 1. A comparison of the ^{31}P resonance assignments for the Dickerson base-paired GC-12- mer, d(CGCGAATTCGCG)$_2$, 1, and the GT mismatch 12-mer, 2 (the GC-12-mer assignments are from Eckstein and coworkers [49]) is also shown in Figs. 7A/B.

In general the ^{31}P chemical shifts of those phosphates in GT-12-mer, 2, furthest from the mismatch sites at position 3 from both 5' ends of the duplex are quite similar to those of the GC-12-mer, 2. The two signals shifted the most upfield (GpT phosphate at position 2) and downfield (TpG phosphate at position 3) occur at the site of mismatch.

Large perturbations in ^{31}P chemical shift are also found in oligonucleotide duplexes possessing a base insert [56]. Thus as shown in Fig. 9 a single phosphate signal is shifted ~ 0.4-0.8 ppm downfield from the main cluster of signals in the 13-mer, 2'.

dCGCAGAATTC GCG
dGCG CTTAAGACGC
2'

^{17}O-labelling of the 13-mer at the phosphate on the 3'-side of the adenosine base insert (Fig. 9B) demonstrates that the unusual downfield shifted resonance belongs to the phosphate involved in the non-Watson-Crick part of the structure. The phosphate adjacent to the base insert on the 5'-side is not significantly perturbed (Fig. 9C).

The ^{17}O labelling scheme provides an important methodology for ^{1}H NMR signal identification. ^{1}H assignments can be made from ^{31}P signals (unambiguously assigned by ^{17}O labelling). As described above 2D NMR provides a mean to identify the base and most of the sugar ^{1}H NMR signals (Figs. 1-3). These sequential resonance assignments via 2D NOESY and COSY, however, require knowing the gross features of the structure in advance (i.e. using the X-ray structure of a B-DNA double helix). In contrast, as described below we have been able to unambiguously make some of the ^{1}H signal identifications by 2D NMR techniques without recourse to any assumed initial geometry [29,35-38,50,51]. This is particularly important if the double helix possesses unusual geometry.

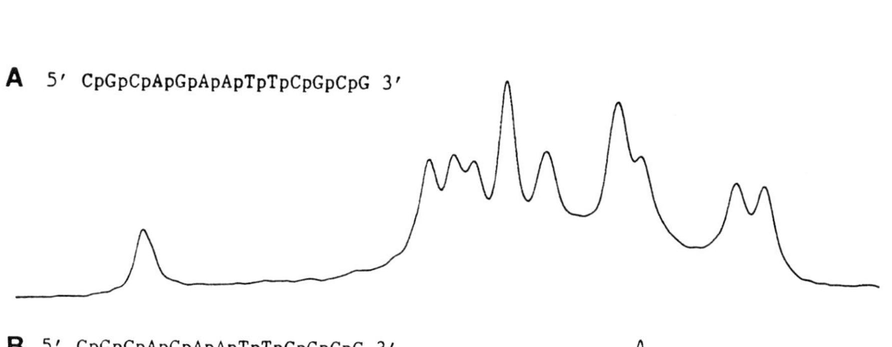

Fig. 9 ^{31}P NMR spectra and phosphate assignments of (A) 13-mer 2'. Examples of site-specific ^{17}O labelling of two of the phosphates of 13-mer 2' at position 4 and 3 (shown by an asterisk) are shown in (B) and (C) respectively.

6.3.2 PAC 2D $^{31}P/^{1}H$ Heteronuclear Correlated Spectra of Oligonucleotides

While the most straightforward and unambiguous method for complete assignment of ^{31}P signals in oligonucleotides relies on site-specific labelling of the monoesterified phosphoryl oxygens of the backbone with ^{17}O or $^{17}O/^{18}O$, [37,38,54,55] it suffers by being rather expensive and time consuming. For an oligomer n nucleotides in length, n – 1 separate syntheses of ^{17}O labelled oligonucleotide are required. 2D ^{31}P-^{1}H NMR spectroscopy [9,52,57,58] can potentially provide a convenient, inexpensive alternative for the assignment of ^{31}P chemical shifts in moderately sized oligonucleotides.

Conventional 2D ^{31}P-^{1}H heteronuclear correlation (HETCOR) NMR spectroscopy has been applied with limited success to short oligonucleotides [20,36-38,50,51,57]. More recently a ^{31}P-^{1}H reverse detection (^{1}H detection) HETCOR experiment has been successfully used to assign ^{31}P signals in a dodecamer duplex [58] and hexamer duplex [59]. For oligonucleotides longer than 4-5 nucleotide residues the simple heteronuclear correlation measurements suffer from poor sensitivity as well as poor resolution in both the ^{1}H and ^{31}P dimensions (particularly at 4.7 Tesla, at which our measurements were made; [20,50,51]). The poor sensitivity is largely due to the fact the ^{1}H-^{31}P scalar coupling constants are about the same size or smaller than the ^{1}H-^{1}H coupling constants.

In agreement with previous reports, [60,61] we have found that sensitivity is substantially improved by using a heteronuclear version of the "constant time" coherence transfer technique, referred to as COLOC (COrrelation spectroscopy via LOng range Coupling) and originally proposed for ^{13}C-^{1}H correlations [61]. We have recently developed a Pure Absorption phase Constant time (PAC) pulse sequence to emphasize ^{1}H-^{31}P correlations in oligonucleotides as well as give rise to pure absorption phase spectra, thereby further improving resolution [9,52]. This pulse sequence incorporated both evolution of antiphase magnetization and chemical shift labelling in a single, "constant time" delay period, thereby improving the efficiency of coherence transfer and increasing sensitivity. The PAC sequence also gives homonuclear decoupling during t_1 and thus improves the resolution.

Superficially, the idea of producing pure absorption phase spectra using a "constant time" sequence seems quite discouraging. In conventional coherence transfer experiments the first order phase correction needed in the t_1 dimension is normally small or even zero. This is due to the fact that the chemical shift labelling delay (t_1) for the first spectrum in the data field is small or zero. In contrast, the first spectrum in a PAC sequence is obtained at a large value of t_1. Consequently, the first order phase correction needed in the t_1 dimension is large, viz., often is many thousands of degrees. However, while this first-order phase correction is extremely large, it can be estimated rather accurately. In a PAC experiment the total first order phase correction can be estimated as (spectral window in ω_1) $\times$ (constant delay) $\times$ 360°, assuming the entire constant delay is used for chemical shift labelling. This is the total first order phase correction from one end of the ω_1 dimension to the other. A more accurate estimate could be made by including the finite lengths of the pulses or these smaller phase corrections can be made interactively. The PAC spectrum of the self-complementary 14 base-pair oligonucleotide d(TGTGAGCGCTCACA)$_2$ is shown in Fig. 10.

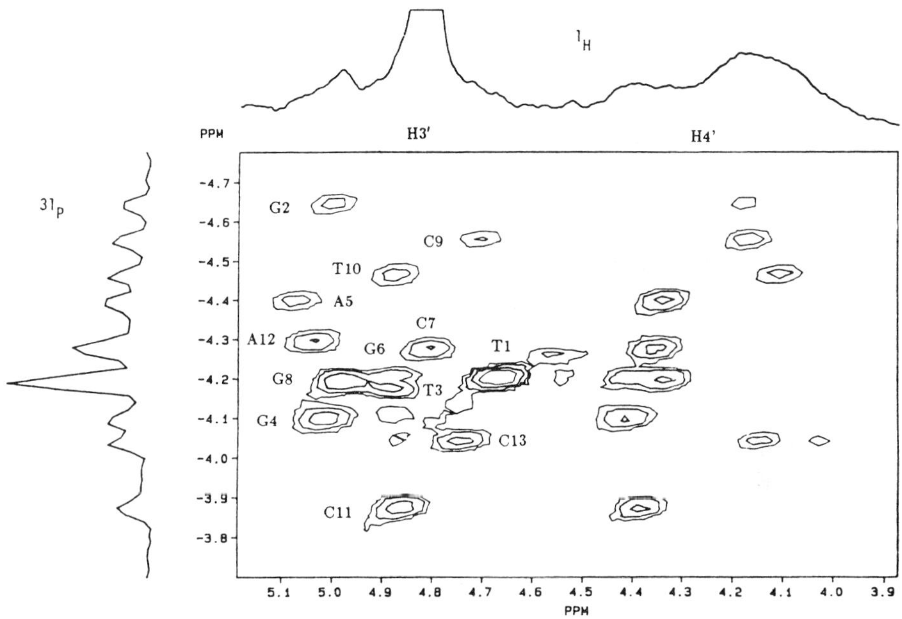

Fig. 10 Pure absorption phase ^{31}P-^1H PAC spectrum of d(TGTGAGCGCTCACA)$_2$ at 200 MHz (^1H). The pulse sequence and phase used for this spectrum was:

PA-90° ^1H(x)-(CD-t_1/2)-composite 180° -(t_1/2)-90° ^1H(ϕ_1)90° ^{31}P(x)-(RD/2)-composite 180° -(RD/2)-t_2(ϕ_2),^1H decouple

^1H: PA 90° CD—t_1/2 180° t_1/2 90° RD/2 180° RD/2 Decouple

^{31}P: PA CD—t_1/2 180° t_1/2 90° RD/2 180° RD/2 t_2

where the composite 180° pulse was a 90°, 180°, 90° sequence with phases x, y, x applied simultaneously to ^1H and ^{31}P resonances, the preacquisition delay PA was 2 s, the constant delay, CD, was 0.051 s, and the refocussing delay RD was 0.035 s. For one of the two states data fields $\phi_1 = \phi_2$=x, −x, x, −x and for the second data field ϕ_1 = y, −y, y, −y and ϕ_2 = −y, y, −y, y. A first order phase correction of 11,520° was used in the ω_1 dimension. The spectrum was acquired on a Varian XL200A spectrometer (200 MHz ^1H).

The cross-peaks represent scalar couplings between ^{31}P nuclei of the backbone and the H3' and H4' deoxyribose protons. The chemical shifts of these protons were determined, as described above, by the sequential assignment methodology for B-DNA using the ^1H/^1H NOESY and COSY spectra. The resulting ^1H chemical shifts and ^{31}P resonance assignments are listed in Table 1.

Tab. 1 ^{1}H and ^{31}P assignments of d(TGTGAGCGCTCACA)2 (*3*) obtained from both 2D ^{1}H-^{1}H sequential assignment methodology, ^{31}P-^{1}H PAC spectrum and ^{17}O labelling of phosphates

Residue	H3' [a]	H4' [a]	^{31}P [b]
	Chemical Shift (ppm)		
T1	4.63 (4.63)	4.00	– 4.21 (– 4.23)
G2	4.95 (4.95)	4.35 (4.37)	– 4.65 (– 4.63)
T3	4.82 (4.84)	4.15 (4.12)	– 4.18 (– 4.19)
G4	4.96 (4.98)	4.28	– 4.11 (– 4.14)
A5	5.01 (5.03)	4.37 (4.37)	– 4.41 (– 4.37)
G6	4.91 (4.84)	4.30 (4.29)	– 4.21 (– 4.19)
C7	4.77 (4.75)	4.30 (4.29)	– 4.28 (– 4.27)
G8	4.93 (4.93)	4.31 (4.30)	– 4.20 (– 4.19)
C9	4.66 (4.66)	4.17	– 4.56 (– 4.54)
T10	4.82 (4.83)	4.13 (4.12)	– 4.48 (– 4.42)
C11	4.82 (4.82)	4.08 (4.06)	– 3.89 (– 3.97)
A12	4.98 (5.00)	4.34 (4.33)	– 4.30 (– 4.23)
C13	4.70 (4.70)	4.06	– 4.05 (– 4.09)
A14	4.61	4.11 (4.10)	

[a] Obtained from 2D ^{1}H-^{1}H sequential assignment methodology [20]. Numbers in parentheses are ^{1}H chemical shifts determined from 2D ^{1}H-^{31}P PAC spectrum and are corrected: actual –0.03 ppm.

[b] From external trimethyl phosphate. Numbers in parentheses are 1D ^{31}P NMR spectrum, assigned via ^{17}O labelling.

Three residues (T3, T10 and C11), however, possess 3' deoxyribose protons with identical chemical shifts, which rendered unambiguous assignment of their corresponding ^{31}P signals impossible on this basis alone. In these cases, assignment of the ^{31}P signal of the ith phosphate was achieved through connectivities with both the 3'H(i) and 4'H(i+l) deoxyribose protons. Although the 5'H(i+l) and 5"H(i+l) protons overlap with the 4' protons, the intensities for the ^{31}P-5' and 5"H PAC cross-peaks generally appear to be much weaker than the 4'H cross-peaks. Similar behaviour has been noted in ^{31}P/^{1}H HETCOR studies on oligonucleotides [50,51,57-58] and presumably reflects the strong and large 5' and 5" coupling. Variations in 3'H and 4'H coupling constants also explain the weaker intensities for some of these PAC cross-peaks.

The ^{31}P resonance assignments have also been determined by the site specific ^{17}O labelling methodology described above. The ^{31}P chemical shift assignments made *independently* by these two methods were identical, confirming both the original ^{31}P assignments by ^{17}O labelling and the utility of the PAC assignment methodology.

In this fashion the ^{31}P chemical shifts of the other oligonucleotides 4-7 and d(GGGCATGCCC)$_2$ (11; see below) were also determined through application of the PAC experiment. (^{1}H chemical shifts were also assigned by the 2D ^{1}H/^{1}H

sequential assignment methodology). The ^{31}P signal assignments for these oligonucleotides were thus completed without recourse to any ^{17}O-labelling and suggests the generality of the PAC methodology. We should emphasize again that sufficient resolution and sensitivity is achievable at 200 MHz (^{1}H) to allow the assignment of the ^{31}P signals of a fairly large biomolecule (MW > 9000 Daltons). In contrast to the rather demanding spectrometer and probe performance required for the reverse detection HETCOR experiment at 500 MHz [57], we demonstrate that PAC can be readily implemented on most commercial, medium-field spectrometers.

6.3.3 DOC 2D $^{31}P/^{1}H$ Heteronuclear Correlated Spectra of Oligonucleotides

Even with the development of the PAC and reverse detection heteronuclear correlation 2D NMR methodologies, ^{31}P resonance assignments via heteronuclear correlation in duplex oligonucleotides represent a difficult 2D NMR problem largely because the resolution is often mediocre in both the proton and ^{31}P spectra, especially at lower fields [9,52,62]. While the PAC method certainly works, the resolution in the proton dimension is digitally limited to the inverse of the constant time delay. Acceptable values of this delay are sharply constrained by the evolution of proton proton antiphase magnetization. This is illustrated by Fig. 11A which shows a series of ^{31}P PAC spectra for the duplex oligomer $d(TGTGAGCGCTCACA)_2$ produced by coherence transfer from protons; the spectra were taken as a function of total evolution time.

As noted above, the "constant time" PAC coherence transfer pulse sequences minimizes this delay by using the same "constant time" for chemical shift labelling and the evolution of heteronuclear antiphase magnetization. This pulse sequence reduces the problem by limiting the time for the evolution of the parasitic proton proton antiphase magnetization. While PAC is a significant improvement over the normal HETCOR experiment, the sensitivity is still significantly reduced by the evolution of the parasitic proton proton antiphase magnetizations.

We have recently proposed a DOuble Constant time (DOC) sequence which addresses the problem more directly by refocusing the proton proton antiphase magnetization with a selective proton 180° pulse in between the two constant delays [63]. Nonselective 180° pulses move through each constant time period to chemical shift label the proton coherences. Thus the proton magnetization which

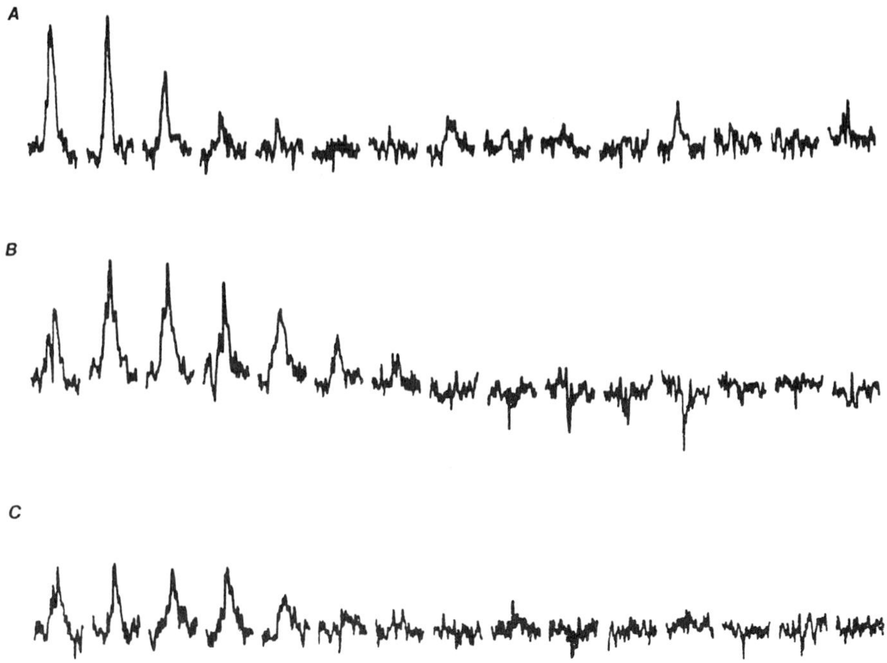

Fig. 11 Comparison of coherence transfer optimal for COLOC (PAC) and DOC. A series of ^{31}P, 1D, coherence transfer spectra of duplex oligonucleotide d(TpGpTpGpApCpGpCpGpTpCpApCpA)$_2$ obtained with the COLOC/PAC and DOC sequences with zero chemical shift evolution. The spectra were measured as a function of the length of the constant time delay. The values of the constant time delay ranged from 20 to 300 ms in 20 ms steps (left to right). The whole ^{31}P spectrum is shown for each time point and a 2 Hz exponential line broadening was used; the individual ^{31}P peaks are not resolved under these conditions. The top series (A) was obtained with the COLOC/PAC sequence, the middle series (B) with the DOC sequence using parameters which gave only C3'-H to ^{31}P coherence transfer, and the bottom series (C) with DOC using parameters which gave C4'-H and C5', 5''-H to ^{31}P coherence transfers. Clearly, with the DOC sequence coherence transfer can be seen for much larger values of the constant time parameter. This is due to the reduction or elimination of ^{1}H-^{1}H coherence transfer. *Reprinted from Ref. 9.*

is antiphase with respect to the dilute spin evolves for the entire constant delay period for protons in the region inverted by the semi-selective 180° pulse. Likewise, the entire constant delay period is effectively used for chemical shift labelling. Proton proton antiphase magnetization which evolves in the first constant time period is refocused in the second constant time period, in the case of interest, i.e. where one proton is within the region inverted by the semi-selective 180° pulse and the proton is outside of this region. This retains the sensitivity and resolution advantages of constant time pulse sequences and provides selectivity in the type of antiphase magnetization present at the end of the two constant time periods.

The DOC sequence is:

^{1}H:	90°	CD– $t_1/4$	180°	$t_1/4$	s180°	CD– $t_1/4$	180°	$t_1/4$	90°	RD/2	s180°	RD/2	dec
X:			180°		180°		180°		90°		180°		AC

where s180° is a semiselective 180° pulse for protons within a selected region; CD is the length of one constant time delay so 2*CD is the total evolution time; RD is a delay for refocusing the heteronuclear antiphase magnetization; t_1 is the delay which starts at 0 and is incremented up to 2*CD to chemical shift label the proton coherences.

Fig. 11A was obtained using the PAC sequence and the rest were obtained using the DOC sequence with different values of the centre frequency of the semi-selective proton 180° pulse. We have found in every case that the acceptable values of the total evolution time are twice as large for DOC as compared to COLOC [63]. This illustrates the well known fact that proton proton antiphase evolution is often the major limitation in obtaining the highest possible resolution in the proton dimension and that DOC has largely or completely eliminated this limitation [52]. This should hold true up to the point where the proton T_2 drops to a value comparable to the time required to evolve significant proton proton antiphase magnetization. The advantages of DOC should also be seen for proton-carbon and other proton-dilute spin heteronuclear correlations.

Fig. 12 shows two DOC spectra differing in their total evolution time.

The first was obtained with a value of the total evolution time (2*CD in the DOC sequence) which we had previously found to be optimal for COLOC spectra. The second was obtained at twice that delay. The much larger delay was made possible by the selectivity of the DOC sequence. It is quite clear that, as expected, twice the resolution is obtained in the proton dimension.

Another important feature is illustrated by the two peaks in the upper left hand corner of the spectrum. These two peaks are much more intense in Fig. 12B than in A. This improvement indicates a substantial increase in coherence transfer efficiency for these peaks as compared to the other peaks. This dramatic improvement most likely indicates that the intensity of these peaks is governed by a smaller value of the H3'-^{31}P scalar coupling. Thus DOC not only can give much better resolution but also *better* sensitivity for the most difficult correlations. This sequence can thus compete effectively with inverse detection schemes, [58] especially where the relative values of the gyromagnetic ratios of the X and ^{1}H spins are not greatly different (as is true for ^{31}P spins). The major limitation of the DOC sequence is that it may be necessary to collect the spectrum in parts (the spectra in Fig. 12 were collected with two separate settings of the center frequency of the semi-selective 180° pulse) and depending on the resolution required in the proton dimension this can take more instrument time. However, if the highest resolution and the observations of the maximum number of correlations is required then the DOC sequence is most likely the best choice.

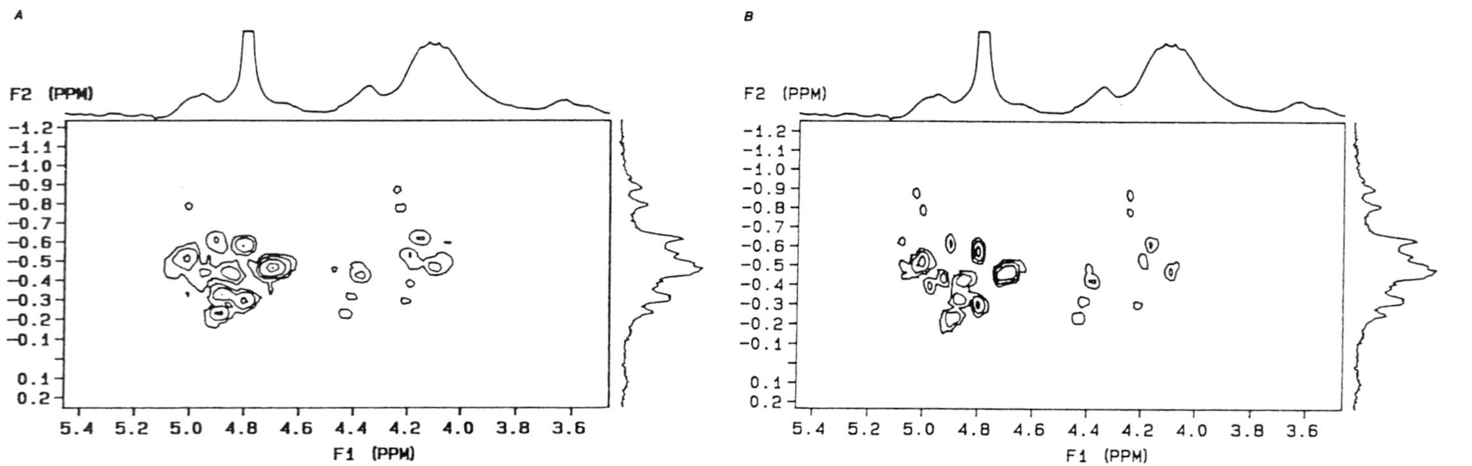

Fig.12 Proton Resolution of DOC Spectra. Two DOC spectra are shown. The first (A) was obtained under conditions which were found to be optimal for COLOC (PAC) spectra, ie. with a constant time delay of 51 ms. The second (B) was obtained with twice the value of the constant delay, 102 ms. The longer delay was made possible by the selective nature of the DOC sequence. For the purpose of comparison both spectra were processed in the same way; resolution enhanced in the ^{31}P (F_2) dimension but not in the ^{1}H (F_1) dimension. The selective ^{1}H 180° pulse with a 20 pulse Dante train that had a total length of 6 ms. Each spectrum was obtained in two parts. For the first part the selective 180° pulse was centred in the region of C3'-H resonances and only C3'-H to ^{31}P coherence transfer peaks were obtained (same setting and selectivity as Fig. 1B). The second part of the spectrum was obtained with the selective 180° pulse centred in the C4'-H and C5', 5''-H region (same as Fig. 1C). After referencing, the second part of the spectrum was plotted on top of the first part. The size of the data sets were 128 data points by 64 FID's for (A) and 128 data points by 128 FID's for (B). The recycle time was 2.4 s and 64 scans were taken per FID. The refocusing delay RD was 40 ms. Rf phase shifts used for the ^{1}H pulses were: 90° pulse ϕ = x, non-selective 180° ϕ = x, semi-selective 180° pulse ϕ = xxxx yyyy –x–x–x–x –y–y–y–y, non-selective 180° pulse ϕ = x, 90° pulse ϕ = xy–x–y, semi-selective 180° pulse ϕ = x. The phase shifts for ^{31}P were: 180° pulse ϕ = x, 180° pulse ϕ = xxxx yyyy –x–x–x–x –y–y–y–y, 180° pulse ϕ = x, 90° ϕ = x, 180° pulse ϕ = x and for the receiver ϕ = x–y–xy –xyx–y. All spectra were recorded on a Varian XL200A spectrometer. *Reprinted from Ref. 9.*

6.3.4 Variation of ^{31}P Chemical Shifts in Oligonucleotides

As described above, earlier ^{31}P NMR studies on poly- and oligonucleic acids supported our suggestion that the base stacked, helical structure with a *gauche, gauche* phosphate ester torsional conformation should be upfield from the random coil conformation. With the ability to identify the individual ^{31}P resonances of oligonucleotides using ^{17}O-labelling or 2D heteronuclear correlation schemes, we can now begin to sort out the various factors responsible for variations in the ^{31}P chemical shifts of oligonucleotides [9,20,29,35-38,50-52,57,64-67]. As discussed above, one of the major contributing factor that we have hypothesized determines ^{31}P chemical shifts is the main chain torsional angles of the individual phosphodiester groups along the oligonucleotide double helix. Phosphates located towards the middle of a B-DNA double helix assume the lower energy, stereoelectronically favoured g^-, g^- and g^-, t conformations, where increased flexibility of the helix is more likely to occur. (The notation for the P-O ester torsion angles follows the convention of Seeman [68] with the ζ, P-O-3' angle given first followed by the α, P-O-5' angle (see Fig. 13).

Because the g^-, g^- conformation is responsible for a more upfield ^{31}P chemical shift, while a g^-, t conformation is associated with a lower field chemical shift, internal phosphates in oligonucleotides would be expected to be upfield of those nearer the ends. Although several exceptions have been observed, this positional relationship appears to be generally valid for oligonucleotides where ^{31}P chemical shift assignments have been determined [9,20,39-41,45,49,50-52,54,65].

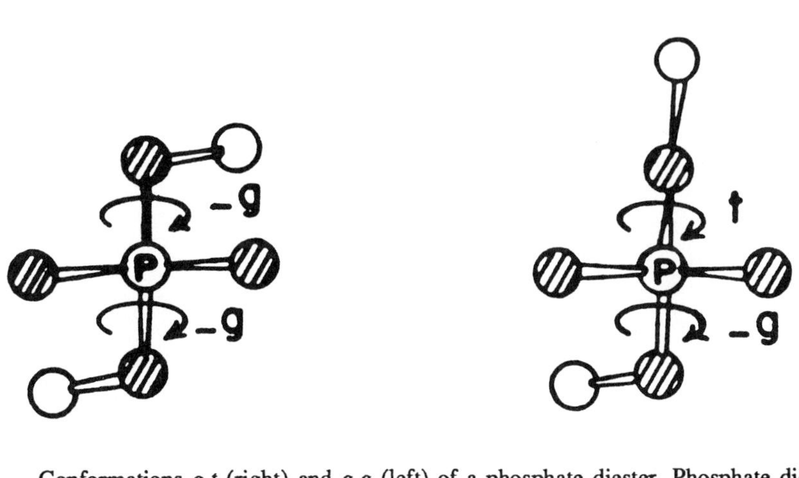

Fig. 13 Conformations *g,t* (right) and *g,g* (left) of a phosphate diester. Phosphate diester torsional angles about the R-O_1-P and P-O_2-(R') bonds are defined by the R-O_1-P-O_2(R') structural fragment and are gauche (60°) or trans (180°). *Reprinted from Ref. 9.*

Thus position of the phosphorus (terminal vs. internal) within the oligonucleotide is one important factor responsible for variations in ^{31}P chemical shifts.

Lerner and Kearns [69], however, have shown that ^{31}P chemical shifts of phosphate esters are modestly sensitive to solvation effects, and have argued that differential solvation of the nucleic acid phosphates could be responsible for the observed variation in the ^{31}P chemical shifts of nucleic acids. Costello et al. [70] have also noted similar sensitivity of ^{31}P chemical shifts to salt. As argued in this paper, it appears as though these environmental effects on the ^{31}P chemical shifts of nucleic acids are smaller than the intrinsic conformational factors discussed above, assuming comparisons are made under similar solvation conditions. Similarly, other possible effects, such as ring-current shifts, are not likely to be responsible for shifts > 0.01 ppm [35,71]. The latter effects are so small because the phosphates are so far removed from the bases (> 10 Å). Bond angle distortions can have a large effect on ^{31}P chemical shifts [35,72-74] but this geometric parameter is coupled to the torsional angle changes so it is not an independent variable.

6.3.5 Sequence-Specific Variation in ^{31}P Chemical Shifts, Calladine Rules

To date, nearly a dozen "modest" sized oligonucleotide sequences have had individual ^{31}P resonances completely assigned. All are, self-complementary (or nearly self-complementary) sequences:

1–7
dGACGATATCGTC
dCTGCTATAGCAG
8
dCATG---G---AT(5-MeC)CATG
dGTAC(5-MeC)TA---G---GTAC
9
dGGAATTCC
dCCTTAAGG
10
dGGGCATGCCC
dCCCGTACGGG
11

As noted in Figs. 7 and 14 ^{31}P chemical shifts of individual phosphates for similar sequences are quite similar. Perturbations in the shifts are largely localized near the site of base-pair substitution.

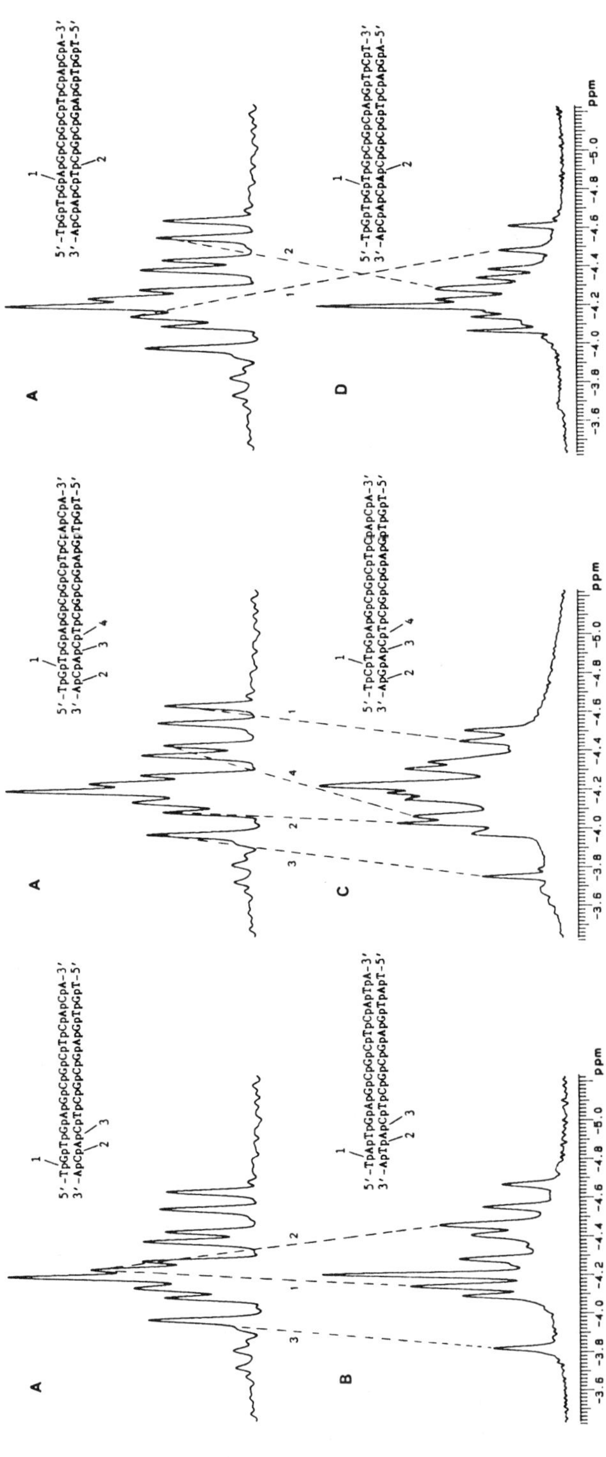

Fig. 14 ^{31}P NMR spectra of 14-mer 1 (B) 14-mer 2, (C) 14-mer 3 and (D) 14-mer 4. Samples contained 4-6 mg purified 14-mers in 0.4 ml D_2O containing 25 mM Hepes, 10 mM EDTA, 75 mM KCL, and 0.1 mM NaN_3, pH* 8.0. *Reprinted from Ref. 9.*

The individual ^{31}P chemical shifts of the oligonucleotide duplexes (referenced to trimethyl phosphate, 0 ppm) are plotted vs. sequence in Figs. 15 and 16.

As shown in Figs. 15 and 16, with the exception of 8-mer, 10, the ^{31}P chemical shifts of the duplex oligonucleotides do not appear to uniformly follow the positional relationship. As would be expected from the phosphate positional relationship, the ^{31}P chemical shifts in the central regions should gradually increase, reach a maximum at the central phosphate position, and decrease once again, due to the favourable energetic considerations of the internucleotidic linkages for internal regions of the oligonucleotide. The central region of the sequences GT-12-mer 2, GC-12-mer 1, 12-mer 8 and 8-mer 10 sequences, appear to follow the positional relationship. However, major variations which do not follow the positional relationship of ^{31}P chemical shifts are observed throughout the 12-mers and 14-mers.

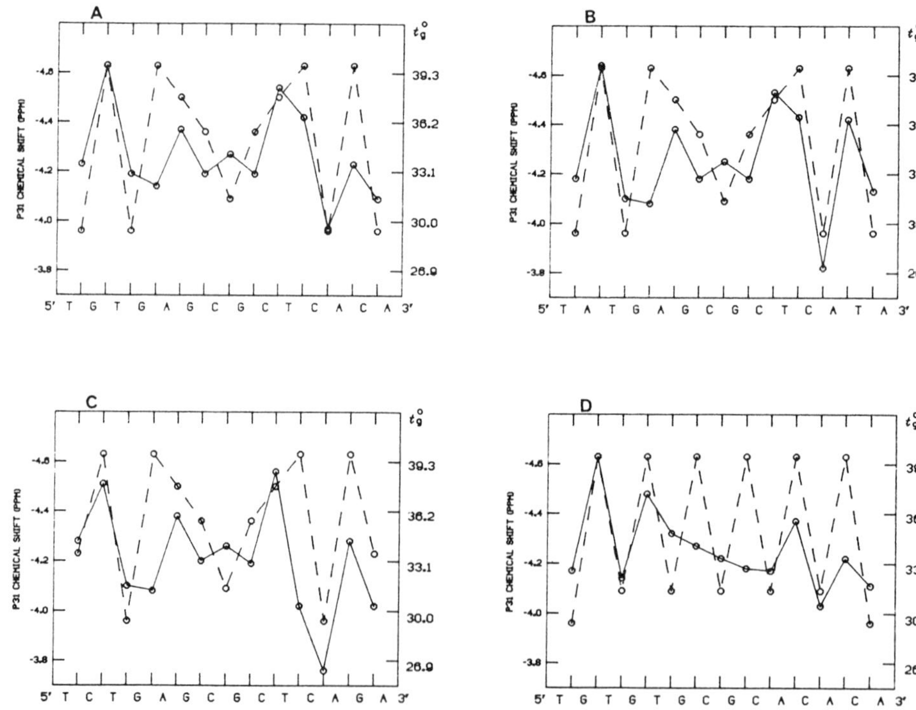

Fig. 15 Plot of ^{31}P chemical shift (—o—) vs. phosphate position along the 5'-3' strand for duplex (A) 14-mer 3, (B) 14-mer 4, (C) 14-mer 5 and (D) 14-mer 6. Also shown is a plot of calculated helix twist sum, t_g, derived from calculated Σ_1 sum function and eqn. (1) ($t_g = 35.6 + 2.1\Sigma_1$) vs. phosphate position (--o--). The t_g vs. sequence plot has been scaled to reflect the ^{31}P chemical shift variations.

As first suggested by Eckstein and coworkers, the ^{31}P chemical shifts appear to vary in response to local, sequence-specific distortions in the duplex geometry. The possible basis for the correlation between sequence-specific helical distortion and ^{31}P chemical shifts can be understood from the following geometrical analysis. Calladine [75] and Dickerson [3,31,32] analyzed the crystal structure of 12-mer 1 and suggested a simple mechanical explanation for the significant

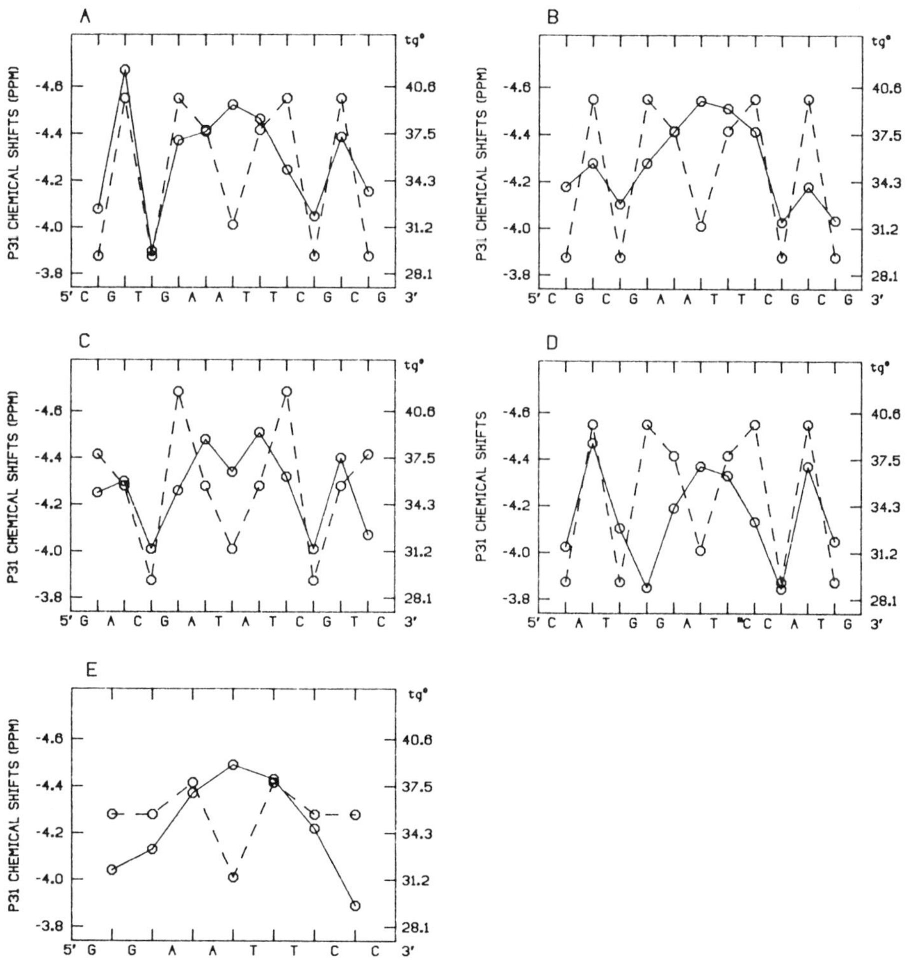

Fig. 16 Plot of ^{31}P chemical shift (—o—) vs. phosphate position along the 5'-3' strand for duplex (A) GT-12-mer 2, (B) GC-12-mer 1, (C) 12-mer 8, (D) 12-mer 9 and (E) 8-mer 10. Also shown is a plot of calculated helix twist sum, t_g, derived from calculated Σ_1 sum function and eqn. (1) ($t_g = 35.6 + 2.1\Sigma_1$) vs. phosphate position (--o--). The t_g vs. sequence plot has been scaled to reflect the ^{31}P chemical shift variations.

distortions from a regular B-DNA double helical structure. It was noted that in order to increase the stacking overlap along each strand of the double helix, the bases along each strand are propeller twisted by 10-20° relative to the bases on the opposite strand. Because the purines extend beyond the helix axis, in a 5'-purine-pyrimidine-3'(5'-Pu-Py-3') sequence or a 5'-Py-Pu-3' sequence the purines on opposite strands that are separated by one base step, sterically clash. The purine N-2/N-3 steric clash in the minor groove of a 5'-Pu-Py-3' sequence and hence a 5'-Py-Pu-3' sequence produces the largest local geometry changes [2,3,31]. Using a simple elastic beam mechanical model, Calladine [2,75] has proposed 4 different conformational variations that will relieve this steric hindrance: (1) flattening the propeller twist, (2) opening the roll angle, (3) displacing the base pairs, and (4) decreasing the local helix twist (Fig. 6).

By mere coincidence five of the oligonucleotide sequences, CG-12-mer 1, 12-mer 9, and the 14-mers 3, 4 and 7, have identical purine/pyrimidine terminal sequence for the first five base pairs, and thus one would initially expect the ^{31}P chemical shift patterns of each terminal region to be identical. It is important to note that the pattern of ^{31}P chemical shift variations at the ends of the two sequences GT-12-mer 2 and 12-mer 8 are nearly identical to that occurring in the GC-12-mer 1, 12-mer 9 and 14-mers 3, 4, 6 and 7 sequences. This is exactly what is observed in each case, and thus one can conclude that the common variation in ^{31}P chemical shift within the two end regions is likely due to the same type of helical adjustment needed to relieve the purine-purine steric clash. These similarities thus strongly support the hypothesis that ^{31}P chemical shift variations arise from sequence-specific distortions in helix geometry.

Fig. 17 compares ^{31}P chemical shifts for complementary positions for the duplex sequences, where the solid line connects ^{31}P chemical shifts for phosphate positions starting at the 5' end and proceeding in the 3' direction, and the dashed line represents ^{31}P chemical shifts for the complementary 3'-5' strand.

Note that because of the palindromic symmetry of each of the oligonucleotides 1-11, the corresponding phosphates on opposite strands that are related by the two-fold dyad axis of symmetry are chemically and hence magnetically equivalent. However, the "complementary" phosphates (phosphates opposite each other on complementary strands) are chemically and magnetically non-equivalent. Rather surprisingly, the ^{31}P chemical shifts at complementary phosphate positions generally follow the same pattern in both strands of the duplex regardless of base sequence or position, suggesting that the phosphate geometry is nearly the same in complementary positions along both strands. Helical adjustment (through variation in helix twist, roll angle, etc.) need not necessarily affect the deoxyribose phosphate linkage at both complementary base steps in an identical manner. For instance, it is not immediately apparent that a change in helical twist will cause torsional angle changes that are identical in both complementary strands, and it

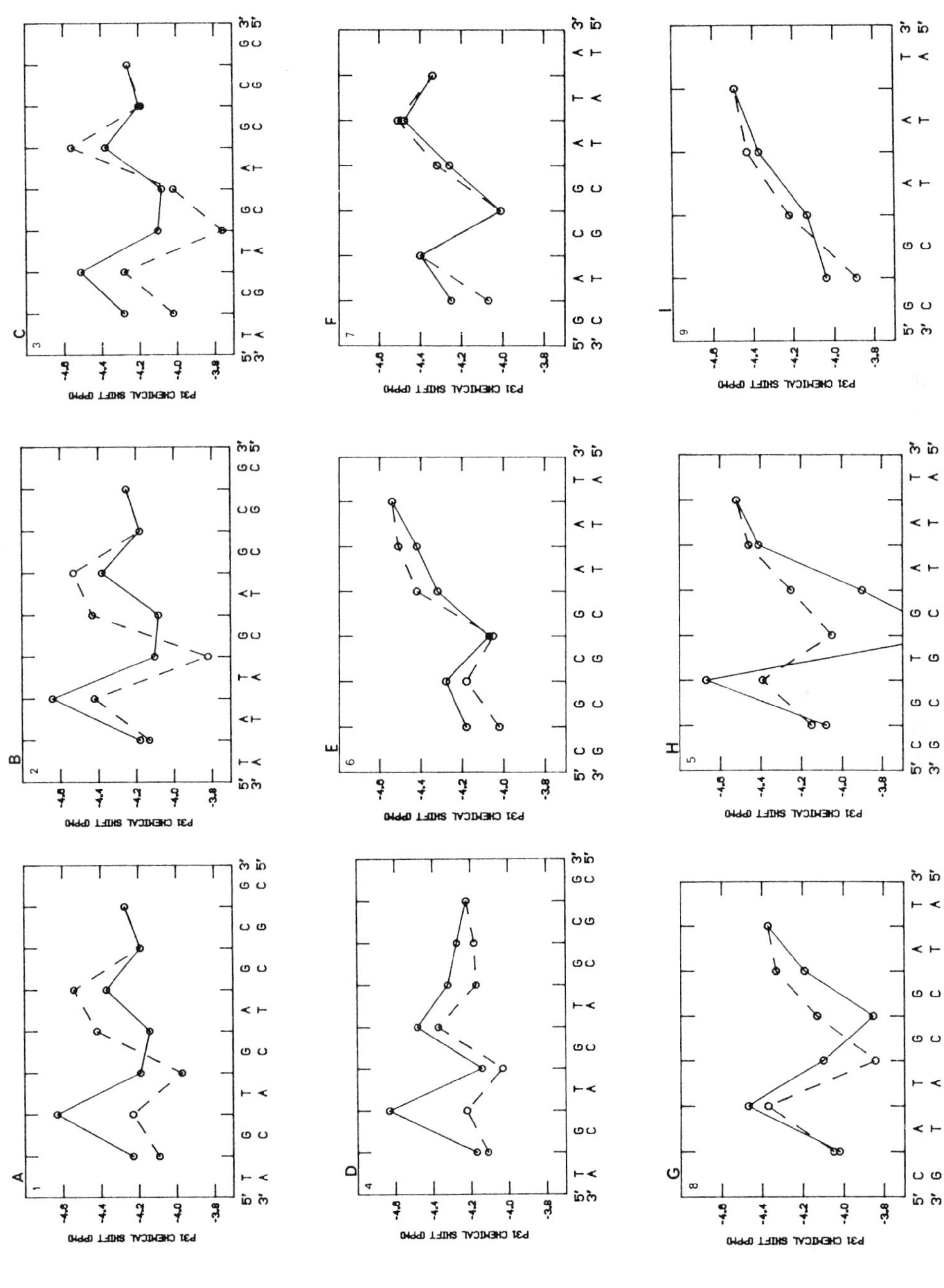

Fig. 17 ^{31}P chemical shift pattern comparison of complementary phosphate positions in oligonucleotides. Because of palindromic symmetry only 1/2 of the sequence is shown. Solid lines: 5'-3' direction; dashed lines complementary 3'-5' direction.

could even be argued that "complementary" changes in torsion angles could even result. In fact, phosphate positions in the 5'-3' strand are always associated with an equivalent or slightly more upfield ^{31}P chemical shifts than the corresponding 3'-5' positions.

These deviations suggest that the base to base interactions and backbone conformations are slightly different along the complementary strands. Similar differences have been observed in the crystal structure of oligonucleotides [2,3,32]. Whereas crystal packing interactions could have been responsible for conformational differences in complementary strands, the fact that we observe similar non-equivalences in the complementary phosphate ^{31}P shifts suggests that these structural differences are intrinsic properties of the "Watson" or "Crick" strands.

As can be seen from the ^{31}P chemical shift pattern of the four phosphate positions shown in Fig. 17, each minor groove clash step at phosphate position 3 is associated with a relative downfield ^{31}P chemical shift, while the adjacent major groove clash step at position 2 results in a large upfield ^{31}P chemical shift. The important point that is seen in Fig. 17 is that while small differences in ^{31}P chemical shifts occur at common purine/pyrimidine steps, the ^{31}P chemical shift variation and absolute chemical shifts of each terminal region phosphate are remarkably similar.

Since the actual base sequence that makes up the terminal regions is different in each of the oligonucleotide sequences 1-11, it would appear that the observed variation of ^{31}P chemical shifts is largely a function only of the purine/pyrimidine sequence. The through space magnetic or electric shielding effects of either a guanosine or adenosine base (i.e. through ring-current and electric charge effects) will not be identical at the phosphorus nucleus. The fact that it is immaterial whether an A or G base is present at a particular purine position in the sequence, further supports the hypothesis that conformational differences arising from purine-purine clash are responsible for the ^{31}P chemical shift variations.

6.3.6 ^{31}P Chemical Shifts and Calladine Rules

As discussed above, Dickerson [32] has shown that the helical distortions observed in the crystal structure of GC-12-mer 1, could be quantitatively predicted through a series of simple "Calladine rule" [2] sum function relationships. Thus the global helical twist (t_g) can be calculated from eqn. 1 using the helical twist sum function (Σ_1). Based upon the ^{31}P assignments of oligonucloetides 1-11, we have suggested that there does appear to be a modest correlation between

^{31}P chemical shifts and the helical twist sum function for a number of oligonucleotides (dashed line in Figs. 15 and 16) [9,50-52]. If we do a least-squares fit of the assigned oligonucleotide ^{31}P shifts to the calculated helical twist (calculated from Σ_1 and eqn. 1), a rather modest correlation appears to exist between the two parameters. The correlation coefficient (R) between ^{31}P shifts and helical twist is 0.71 (Fig. 18) if just the terminal phosphates and the ^{31}P chemical shifts of several drug complexes are included in the correlation. Surprisingly in the middle region, no correlation (R = 0.05) is found between ^{31}P chemical shifts and helix twist (see for example, Figs. 16 and 16B).

Either ^{31}P chemical shifts behave differently in these two regions of the duplexes or the sequence-specific "Calladine rules" derived from the X-ray crys-

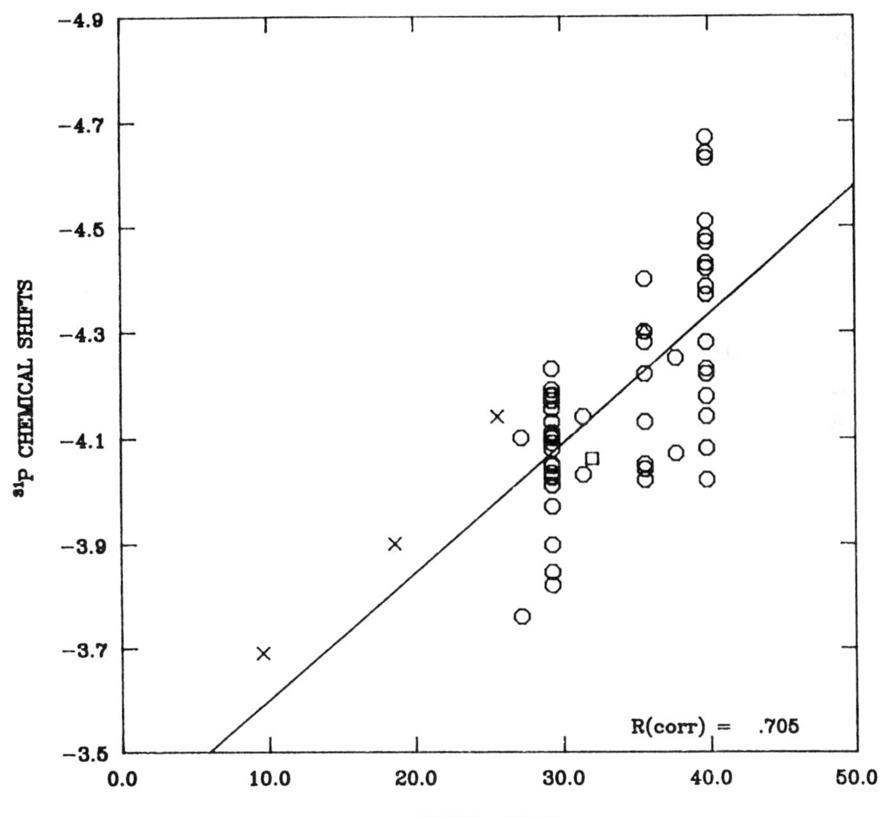

Fig. 18 Plot of ^{31}P chemical shift vs. calculated helix twist, t_g, derived from eqn. (1) ($t_g = 35.6 + 2.1\Sigma_1$). Oligonucleotide 1-11 terminal phosphate shifts (o), drug-DNA (x), B-DNA (Δ) and alternating poly d(AT) (□) shift are shown.

tal structures of oligonucleotides is at least partially invalid. As noted previously, 2D NMR and molecular mechanics and dynamics calculations have also produced mixed results in support of the large sequence-specific variations in duplex geometry that are observed in the solid state.

6.3.7 Origin of ^{31}P Chemical Shift Variations and Calladine Rules

As noted above, the possible basis for the correlation between Calladine rule sum functions and ^{31}P chemical shifts can be analysed in terms of deoxyribose phosphate backbone distortions involved in relief of purine-purine clash. Thus, decreasing the helical twist angle, t_g reduces the steric clashing in the minor groove in a 5'-Py-Pu-3' sequence by pulling the N-2 and N-3 atoms of the purines further apart. As the helix unwinds (and the helix twist t_g decreases), the length of the deoxyribose phosphate backbone might be expected to decrease. These local helical changes require changes in the deoxyribose phosphate backbone angles α-ζ:

$$P^{\alpha} - O5'^{\beta} - C5'^{\gamma} - C4'^{\delta} - C3'^{\varepsilon} - O3'^{\zeta} - P$$

As the helix winds or unwinds, the distance between the adjacent C-4' atoms of deoxyribose rings along an individual strand ($D_{4'4'}$) must change to reflect the stretching and contracting of the deoxyribose phosphate backbone between the two stacked base pairs. To a significant extent, these changes in the overall length of the deoxyribose phosphate backbone "tether" are reflected in changes in the P-O ester (as well as other) torsional angles.

The sequence-specific variations in the P-O (and C-O) torsional angles may provide the linkage between the Calladine-rule-type sequence dependent structural variations in the duplex and ^{31}P chemical shifts.

Analysis of the X-ray crystal structures [1,3,31] of a B-form oligodeoxyribonucleotide (1) has shown that torsional angles α, β and γ on the 5'-side of the sugar are largely constrained to values g^- (– 60°), t (180°), g(+ 60°), whereas significant variations are observed on the 3'-side of the deoxyribose phosphate backbone. The greatest variation in backbone torsional angles is observed for ζ(P-O3') followed by ε(C3'-O3') and then δ(C4'-C3'). It is important to note that many of these torsional angle variations are correlated. Torsional angle δ correlates with the sugar-puckering conformation and the difference in δ between the two ends of a base pair correlates with the Dickerson/Calladine sum function Σ_3. The sugar ring constrains δ to values no smaller than 70°-85° (C3'-endo) and no larger than 140°-160° (C2'-endo). Intermediate values of δ yield the other two common DNA sugar-puckering conformations ($\delta \sim 96°$: O1'-endo and ~ 120°: C1'-exo). As noted by Dickerson δ, ε and ζ are all correlated. At low values of ζ, the phosphate P-O3' conformation is in the low energy g^- conformation. When the P-O3' conformation is g^-, invariably the C-O3' conformation (ε) is found to be t. This ε (t), ζ (g^-) conformation is the most common backbone conformation. In this B_I (t, g) conformation, δ can vary considerably [21,22]. The other most common conformation for the (ε, ς) pair is the (g^-, t) or B_{II} state. A "crankshaft" motion interconverts B_I and B_{II} conformations with remarkably little overall movement of the phosphate. As δ increases to its maximum limit of ~ 160° in order to relieve any additional steric clash, ζ now begins to vary from – 60° to – 120° while ε largely remains fixed at trans. At a maximal value for δ the phosphate ester conformation can switch to the B_{II} state. It may well be significant that only terminal phosphates are observed in the B_{II} conformation in the crystal structure, thus providing a possible rationale for the different helical adjustments that appear to be responsible for the ^{31}P chemical shift variations in different regions of the helix. It is largely this variation in δ, ε and ζ that allows the sugar phosphate backbone to "stretch" or "contract" to allow for variations in the local helical twist and base-pair displacements of B-DNA. Although B-DNA winds at an average of 36°/nucleotide (~ 10 bps/turn), X-ray crystal structures have shown variation between 9 and 12 bps/turn. As shown by Dickerson for phosphates in a B_I conformation, a good correlation exists between δ and the helix rotation angle, τ, defined between adjacent phosphates along a strand. Unwinding the double helix (increasing τ and decreasing the number of base pairs (n) per turn of the helix) requires torsion about δ, ε and/or ζ to open up the deoxyribose phosphate backbone to span the larger separation between stacked base pairs.

6.3.8 Molecular Mechanics Energy Minimization Calculation of the Sequence-Specific Variation in Deoxyribose Phosphate Backbone Conformation

Molecular mechanics and molecular dynamics calculations have not generally been able to reproduce the Calladine-rule type of sequence-specific variation in the duplex geometry [14,16]. We have recently [9] been able to model changes in the structure of a simple duplex dimer as a function of helix twist and reproduce some of the sequence-specific variations in the crystallographic data. The molecular mechanics program AMBER [30] was used to generate a series of idealized Arnott B-DNA dinucleoside monophosphate duplex structures [33] with sequence d(TG) · d(CA) (Fig. 6) in which the helix twist was varied from 25° to 45° while the helical height was kept fixed at 3.38 Å. The model built structures were then energy refined until a rms gradient of 0.1 kcal/mol-Å was achieved or until the change in energy was less than 1.0×10^{-7} kcal/mol for successive steps. The four C4' atoms were constrained in all structures (force constant 9999.99 kcal/mol-$Å^2$) except for a control duplex with model-built helix twist of 36.0°.

In order to properly analyze the backbone changes as a function of a helix twist, we have introduced a new geometric parameter, the distance between adjacent C4' atoms along an individual strand ($D_{C4'C4'}$). This distance appears to more accurately reflect the stretching and contracting of the deoxyribose phosphate backbone between two stacked base pairs. Using the AMBER energy-minimized, constrained d(TpG) · d(CpA) duplex dimer model, we have calculated the variation in $D_{C4'C4'}$ as a function of helix twist (Fig. 19).

Decreasing the twist angle, t_g from 36° to 25° reduces the steric clashing in the minor groove in a 5'-Py-Pu-3' sequence by pulling the N-2 and N-3 atoms of the purines further apart. As the helix unwinds, the length of the deoxyribose phosphate backbone tether is expected to decrease: the calculations suggest a decrease from ~ 5.8 Å to 4.7 Å. As the helix more tightly winds with t_g increasing to 45°, the $D_{C4'C4'}$ increased to 6.7 Å.

Significantly, the $D_{C4'C4'}$ distances obtained from the four crystal structures [1,3,31] of 12-mer oligonucleotide 1 also follow a similar change as a function of t_g, as shown in Fig. 19. The correlation coefficient between the crystallographically-derived $D_{C4'C4'}$ distances and t_g is a quite respectable value of 0.77. While the calculated distances respond more steeply to the variation in t_g than the actual values, the trend is similar. This is reasonable since we are ignoring other helical adjustments (e.g. roll and base slide) which must also perturb the backbone length. It should be noted that only the B_I backbone conformations were included in the data of Fig. 19 and the correlation breaks down if both B_I and B_{II} conformations are analyzed.

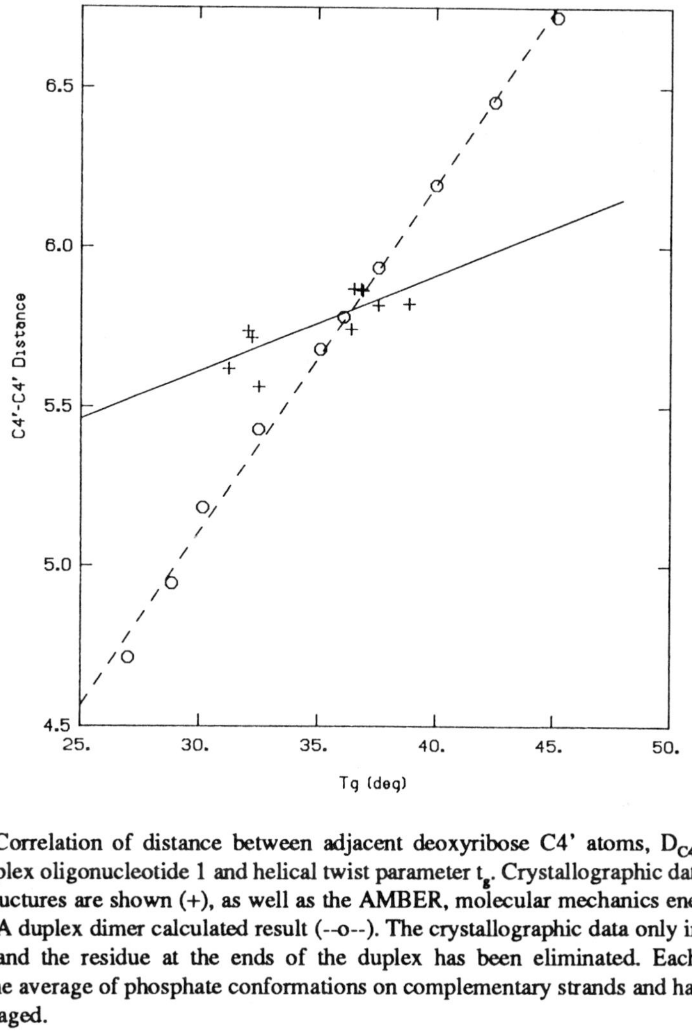

Fig. 19 Correlation of distance between adjacent deoxyribose C4' atoms, $D_{C4'C4'}$, along one strand of duplex oligonucleotide 1 and helical twist parameter t_g. Crystallographic data from the four 12-mer 1 structures are shown (+), as well as the AMBER, molecular mechanics energy minimized dTpG · dCpA duplex dimer calculated result (--o--). The crystallographic data only includes B_I conformations and the residue at the ends of the duplex has been eliminated. Each conformation represents the average of phosphate conformations on complementary strands and has also been end for end averaged.

6.3.9 Sequence-Specific Variation of ^{31}P Chemical Shifts and Backbone Torsional Angles

These conformational changes may provide an explanation for the Calladine-rule-type sequence dependent variation in ^{31}P chemical shifts. As mentioned above, two of the most important parameters controlling ^{31}P chemical shifts in phosphate esters are the P-O torsional angles (in nucleic acids the α and ζ torsional angles) [5,73], and C-O torsional angles (β, ε) [72,76], although the P-O torsional angle may be more important. It is thus most significant that there appears to be a strong correlation between the C3'-O3'-P-O5' (ζ) torsion angles and ^{31}P chemical shifts in the 12-mer 1.

Utilizing a heteronuclear proton-flip or a selective version of the ^{1}H-detected heteronuclear pulse sequence [9,77] a number of the backbone coupling constants (and through the Karplus relationship, the torsional angles) of oligonucleotides 1-11 have now been measured. Using these measured $J_{H3'-P}$ coupling constants and the Karplus relationship, H3'-C3'-O3'-P torsional angles may be calculated. From these values we have calculated both C4'-C3'-O3'-P (ε) and C3'-O3'-P-O5' (ζ) torsional angles. (Recall the strong correlation, R = 0.92, shown by Dickerson [3,20] between torsional angles ζ and ε in the crystal structure of 1. ζ was calculated from the relationship $\zeta = -347 - 1.22\,\varepsilon$). Shown in Figs. 20 and 21 are plots of the variation of both ζ and ^{31}P chemical shifts for a number of the phosphates or oligonucleotides 1-11 for which both ^{31}P chemical shifts and $J_{H3'-P}$ coupling constants have been assigned.

The correlation coefficient between ζ (or ε) and ^{31}P chemical shifts is a very good 0.90 for 12-mer 1 and 0.70 for all of the data in Fig. 21 [9,63].

In those instances where good crystal structures are available (such as 1) it is important to note that significant differences exist between the unrestrained and restrained molecular dynamics calculated structure, the backbone conformation calculated from the measured solution coupling constants and ^{31}P chemical shifts and the solid-state crystal structure. As shown in Fig. 20, $J_{H3'-P}$ (and by inference ζ and ε) follows the helix twist sum function pattern only at the ends of the helix,

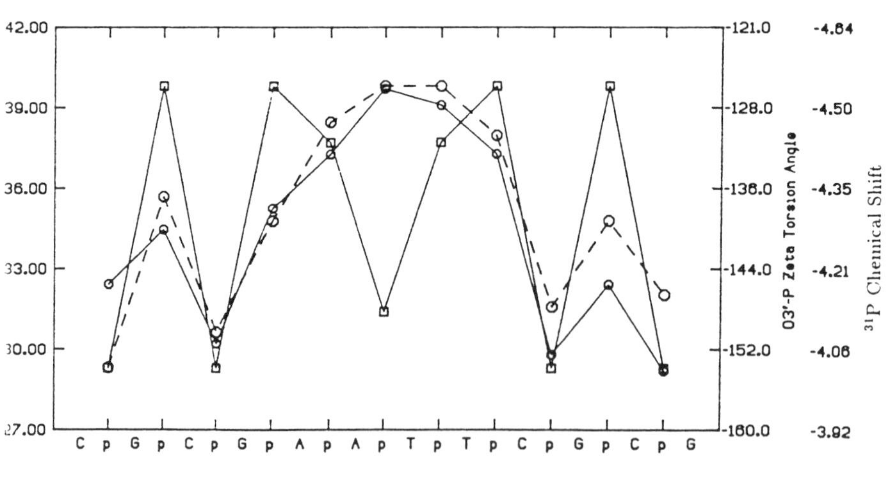

Fig. 20 Comparison of ^{31}P chemical shifts with P-O ester torsional angle ζ for 12-mer 1. Assignments of ^{31}P chemical shifts from Ott and Eckstein [49] and ζ torsional angles calculated from reported $J_{H3'-P}$ coupling constants, the Karplus relationship and the correlation between ε and ζ [9,77].

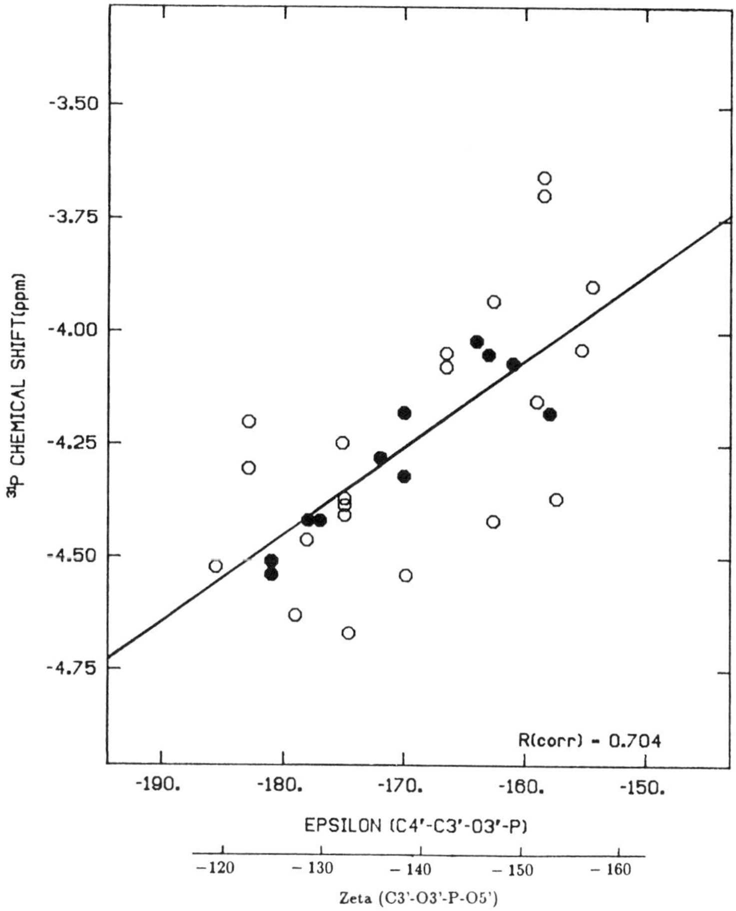

Fig. 21 Plot of ^{31}P chemical shift vs P-O3' ester torsional angle ζ and 3'C-O torsional angle ε for 12-mer 1 (•) and some of the other phosphates of oligonucleotides 2-7; 11 (o). ζ torsional angle calculated from measured $J_{H3'-P}$ coupling constants and the Karplus relationship [34,55].

just as is observed for the ^{31}P chemical shifts (Figs. 15 and 16). This further substantiates the point we earlier made regarding the distinction between helix adjustment in response to steric clash at the ends vs. the middles of the duplex. Dickerson and Calladine have shown that sequence-specific helix distortions as measured by adjustments of the bases are well represented *throughout the entire helix* by the sum function relationships (except for the residue at each end of the helix).

Dickerson has also noted that there appears to be little correlation between the helical parameters and the deoxyribose phosphate backbone torsional angles. However, analysis of the crystallographically derived backbone torsional angles for 12-mer 1 shows a modest (Fig. 22) correlation between t_g and ζ (R = – 0.50), consistent with the AMBER-calculated ζ torsional angle variations (Fig. 22). No significant correlations are observed between α and ε and t_g, although our calculations would suggest a modest variation in ε.

The calculations and data of Figs. 19-22 strongly support our hypothesis that the sequence-specific variation in ^{31}P chemical shifts is largely attributable to sequence-specific variations in the helical parameters and the backbone torsional angles (at least at the ends of the helices for ζ and ε torsional angles). Unwinding the double helix decreases $D_{C4'C4'}$ and it is largely δ, α and ζ that vary in response to variations in t_g and $D_{C4'C4'}$.

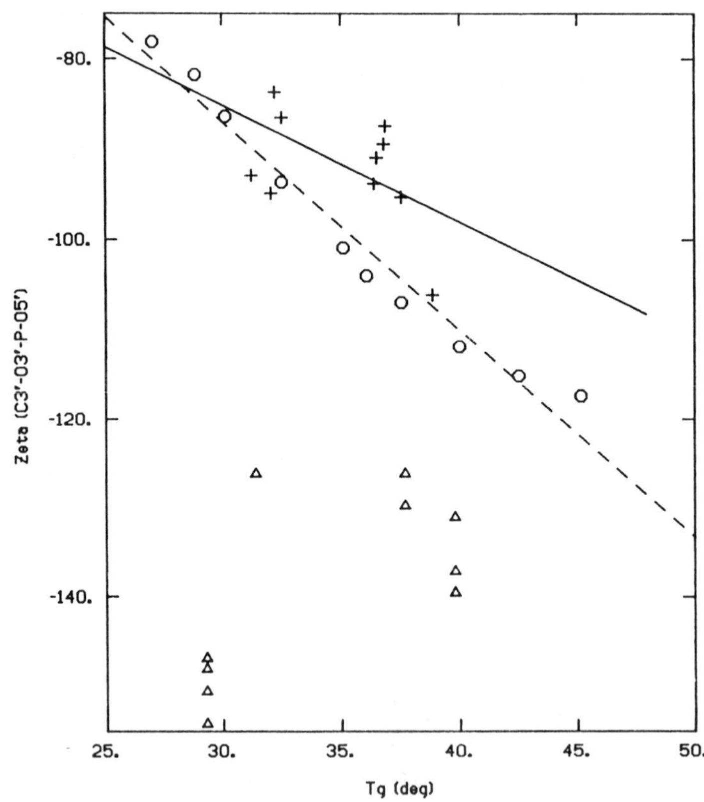

Fig. 22 Plot of calculated ζ, P-O3' torsional angles as a function of helix twist, t_g, based upon the AMBER, molecular mechanics energy minimized dTpG · dCpA duplex dimer structures (--o--). Also shown are the ζ torsional angles derived from the X-ray crystal structures of 12-mer 1 (-+-). The crystallographic data only includes B_I conformations and the residue at the ends of the duplex has been eliminated. Each conformation represents the average of phosphate conformations on complementary strands and has also been end for end averaged. The ζ torsional angles derived from the solution coupling constants are also shown (Δ). Lines represent the best least-squares fit of the crystal and theoretical data.

6.4 Binding of *Lac* Operator to *Lac* Repressor Headpiece Protein

The *lac* repressor system is ideal for studying DNA-protein interaction by NMR [78-81]. It appears to be possible to duplicate the basic *lac* operator *lac* repressor protein interaction by using the smaller *lac* repressor headpiece N-terminal domain fragment [22,82] and *lac* operator DNA fragments such as the 14-mer 3 (two subunits of tetrameric *lac* repressor bind to the two halves of the palindromic lac operator [83]). Kaptein and coworkers [22,84] have assigned many of the ^{1}H signals of the *lac* repressor headpiece by 2D NMR methods. Recent 2D NMR and NOESY distance-restrained molecular dynamics studies of repressor headpiece bound to a *lac* operator DNA fragment have begun to provide details confirming the sequence-specific interactions of a recognition α-helix binding within the major groove of the operator DNA (interestingly, the NMR results are in conflict with models for the 434 and *cro* repressor DNA complexes derived from the X-ray structures [4,84]).

The ^{31}P chemical shift changes upon binding the 56-residue headpiece to the 14-mer 3 are shown in Fig. 23. Only small (< 0.3) ppm perturbations of the ^{31}P chemical shifts are observed in the complex relative to the free duplex and suggest that recognition occurs via electrostatic, groove binding interactions rather than intercalation, since interaction is expected to produce a 2-3 ppm downfield ^{31}P shift while electrostatic association should produce considerably smaller (and generally upfield shifts) of the phosphate directly involved in the interaction [35,36,45-48,67,85,86]. Surprisingly we find that the ^{31}P titration curve levels off at a 1:1 complex (Fig. 23). This suggests that a second headpiece protein is physically blocked from binding to the super-operator. Note also we are in rapid chemical exchange and thus the spectra are still quite simple and easy to interpret. Furthermore the interaction is highly specific with the only major changes occuring at phosphates at positions 2 (upfield) and 3 (downfield) (i.e. TpGp*Tp*G...., asterisks indicating position of interaction). Most importantly, the interaction is asymmetric, with little perturbation of phosphates 11 and 12 on the opposite strand. This is the first direct evidence for this asymmetry and may be an important component of the binding process. (Other ^{1}H NMR studies in the imino region have shown perturbation in the AT* base pair in the TpGpT*pG portion of a symmetrical 18-bp operator 51-residue headpiece complex; [87]).

The ^{31}P NMR data is consistent with current models for the repressor DNA complex (Fig. 24).

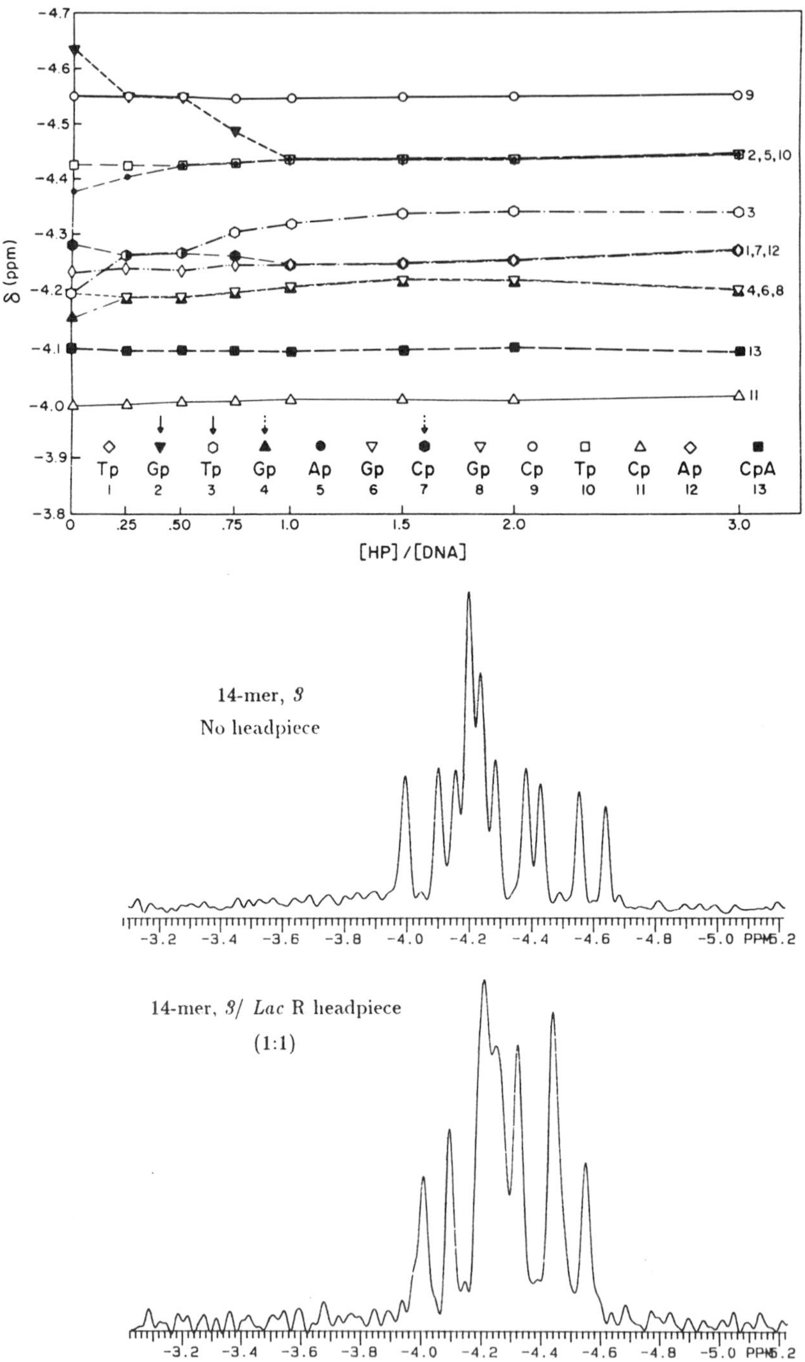

Fig. 23 (A) Titration of the ^{31}P chemical shifts and assignments of the 14 bp oligonucleotide duplex 3 as a function of the relative ratio of [*lac* repressor headpiece]/[14 bp "super" operator, 3] (B) ^{31}P NMR spectra of the 14-mer in the presence of added 56-residue *lac* repressor headpiece.

The above model has been derived structures from the NMR studies of Kaptein and coworkers, as well as related X-ray crystallographic structures for other DNA binding proteins [4,88]. A number of recent studies on site-specific mutates of the *lac* repressor [89-92] also suggest that the major recognition site of the operator is the 5'-TGTGA sequence. Lehming et al.'s [89] data suggest that Tyr17 in the recognition helix interacts with the operator near this site, consistent with our ^{31}P NMR data.

Fig. 24 MIDAS model-built structure of *lac* repressor headpiece complex with operator oligonucleotide fragment.

6.5 DNAase I; Significance of the Local Variation in Phosphate Ester Geometry?

Of potential significance, Klug and coworkers [93] have shown that the sites of DNAase I catalysed hydrolysis of the palindromic 12-mer, d(CGCGAATTCGCG), correlates with the Calladine rules for sequence-specific local variation in duplex geometry. Drew and Travers [94,95] have suggested that these sequence-specific variations in the DNA structure could be responsible for the specificity of DNAase I, and DNAase II, and copper-phenanthroline catalyzed hydrolysis of DNA. Although quite speculative, it may very well be possible that this dependence of the enzymatic (and non-enzymatic, copper-phenanthroline) catalyzed hydrolysis of the DNA to local helical structure [93,94] is a reflection of sequence-specific variation in the P-O ester torsional angles - a stereoelectronic effect on phosphate ester hydrolysis [50,51]. Thus, DNAase I has a very marked reference for hydrolyzing poly d(AT) on the 5' side of T residues [6,96]. Based upon the crystal structure [97] of $(d\text{-}AT)_2$ and the DNAase I catalyzed hydrolysis of poly d(AT), Klug and coworkers [93] suggested that poly d(AT) exists in an "alternating-B" conformation. Significantly the ^{31}P spectrum of poly d(AT) gives two signals separated by as much as 0.8 ppm depending upon salt conditions [98]. By thiophosphoryl labelling, Eckstein et al [65] were able to establish that the downfield shifted ^{31}P signal arises from the TpA phosphates, which based upon an X-ray crystal model, are in a more extended trans-like phosphate ester conformation. The ^{31}P signal of the ApT phosphates is quite similar to that of a normal B-DNA phosphates and indeed is in a g^{-}, g conformation.

It should be noted that poly d(AT) in low salt solution shows a separation of ^{31}P signals of 0.24 ppm [98]. According to our ^{31}P chemical shift/helical twist correlation, this downfield TpA phosphate signal suggests an alternating helical twist of ~ 32° and 36° for the TpA and ApT portions of the duplex, respectively, consistent with the X-ray studies. Similarly, the ^{31}P signal of chicken erythrocyte DNA shifts downfield by 0.2 to 0.3 ppm as it is converted from the B to the A-DNA conformation [99]. A-DNA shows a t_g of ca. 32°-34° in contrast to B-DNA with t_g of ca. 36°. These downfield shifts of the TpA phosphate in poly d(AT) and in A-DNA phosphates are thus in agreement with the stereoelectronic, torsional angle effect on ^{31}P chemical shifts.

6.6 Conclusions

Through 2D NMR or ^{17}O-labelling experiments it is now possible to unambiguously assign the ^{1}H and ^{31}P signals of modest-sized duplex oligonucleotides. The major structural variation responsible for ^{31}P shift perturbation appears to be

P-O and C-O backbone torsional angles which respond to changes in the local helical structure (e.g. helix twist). However, ^{31}P chemical shifts and $J_{H3'-P}$ coupling constants both indicate that these backbone torsional angle variations are more permissive at the ends of the double helix than in the middle and that the solution torsional angles do not always agree with the crystal data. In contrast sequence-specific variations in duplex structures based upon X-ray crystallographic determinations do not show region-specific variations. Similarly, calculated structures in solution based upon NOESY-distance restrained molecular mechanics and dynamics fail to entirely reproduce the "Calladine rule" sequence-specific structural variations. While we have no explanation for these discrepancies at the present time, clearly, ^{1}H and ^{31}P 1D and 2D NMR spectroscopy in conjunction with molecular mechanics and dynamics calculations appear to be able to provide a powerful probe of the conformation of DNA in solution.

Acknowledgements

Supported by NIH (GM36281), the Purdue University Biochemical Magnetic Resonance Laboratory which is supported by NIH (grant RR01077 from the Biotechnology Resources Program of the Division of Research Resources), and the NSF National Biological Facilities Centre on Biomolecular NMR, Structure and Design at Purdue (grants BBS 8614177 and 8714258 from the Division of Biological Instrumentation). The contributions of B. Spangler, D. Shah, K. Lai, P. Wisniowski and A.C. Moulin are much appreciated.

References

[1] Saenger, W. "Principles of Nucleic Acid Structure" Springer-Verlag, New York. (1984).

[2] Calladine, C.R. *J. Mol. Biol.* (1982) *161*, 343-352

[3] Dickerson, R.E. *J. Mol. Biol.* (1983) *166*, 419-441.

[4] Anderson, J.E., Plashne, M. and Harrison, S.C. *Nature*, (1987) *326*, 846-852.

[5] Gorenstein, D.G. *Chem. Rev.* (1987) *87*, 1047-1077.

[6] Klug, A., Viswamitra, M.A., Kennard, O., Shakked, Z. and Steitz, T.A. *J. Mol. Biol.* (1979) *131*, 669-680.

[7] Lefevre, J., Lane, A.N. and Jardetzky, O. *Biochemistry* (1987) *26*, 5076-5090.

[8] Nilges, M., Clore, G.M. Gronenborn, A.M., Piel, N. and McLaughlin, L.W. *Biochemistry* (1987) *26*, 3734-3744.

[9] Gorenstein, D.G., Schroder, S.A., Fu, J.M., Metz, J.T., Roongta, V. and Jones, C. *Biochemistry* (1988) *27*, 7223.

[10] Assa-Munt, N. and Kearns, D.R. *Biochemistry* (1984) *23*, 791.

[11] Clore, G.M., Gronenborn, A.M., Brunger, A.T. and Karplus, M., *Journal of Molecular Biology* (1985) *186*, 435-455

[12] Patel, D.J., Shapiro, L. and Hare, D. *ibid. Ann. Rev. Biophys. Chem.* (1987) *16*, 423-54.

[13] Rinkel, L.J., Van der Marel, G.A., Van Boom, J.H. and Altona, C. *Eur. J. Biochem.* (1987) *163*, 275-286.

[14] Kollman, P., et al. *Biopolymers* (1982) *21*, 2345.

[15] Levitt, M. *Proc. Natl. Acad. Sci.* (1978) *78*, 640-644.

[16] Singh, U.C. *Proc. Natl. Acad. Sci.* (1985) *82*, 755.

[17] Hare, D.R., Wemmer, D.E. Chou, S.H., Drobny, G. and Reid, B. *J. Mol. Biol.* (1983) *171*, 319.

[18] Scheek, R.M., Boelens, R., Russo, N., Van Boom, J.H. and Kaptein, R. *Biochemistry* (1984) *23*, 1371-1376.

[19] Feigon, J., Leupin, W., Denny, W.A. and Kearns, D.R. *Biochemistry* (1983) *22*, 5930-5942, 5943-5951.

[20] Schroeder, S., Fu, J., Jones, C. and Gorenstein, D.G. *Biochemistry* (1987) *26*, 3812-3821.

[21] Kumar, A., Wagner, G., Ernst, R.R. and Wüthrich, K. *J. Am. Chem. Soc.* (1979) *103*, 3645-3658.

[22] Zuiderweg, E.R.P., Scheek, R.M., Boelens, R., Gunstern, W.F.van, Kaptein, R. *Biochemie* (1985) *67*, 707.

[23] Gronenborn, A.M and Clore, G.M. *Progr. NMR Spectrosc.* (1985) *17*, 1.

[24] Keepers, J.W. and James, T.L. *J. Magn. Reson.* (1984) *57*, 404.

[25] Wüthrich, K. "NMR of proteins and Nucleic Acids" John Wiley and Sons, New York, N.Y. (1986).

[26] Clore, G.M. and Gronenborn, A.M. *J. Magn. Reson.* (1983) *53*, 423.

[27] Havel, T.F., Kuntz, I.D. and Crippen, G.M. *Bull. Math. Biol.* (1983) *45*, 665-720.

[28] Braun, W and Go, N. *J. Mol. Bio.* (1985) *186*, 611-626.

[29] Gorenstein, D.G., Schroeder, S.A., Miyasaki, M., Fu, J.M., Roongta, V., Abuaf, P., Metz, J.T. and Jones, C.R. *Bull. Magn. Reson.* (1986) *8*, 137-146.

[30] Weiner, P.K. and Kollman, P.A. *J. Comp. Chem.* (1981) *2*, 287.

[31] Dickerson, R.E. and Drew, H.R. *J. Mol. Biol.* (1983) 149, 761-786.

[32] Fratini, A.V., Kopka, M.L. Drew, H.R. and Dickerson, R.E. *J. Biol. Chem.* (1982) *257*, 14686-14707.

[33] Arnott, S. and Hukins, D.W.L. *Biochem. Biophys. Res. Commun.* (1972) *47*, 1504-1509.

[34] Sundaralingam, M. *Biopolymers* (1969) *7*, 821.

[35] Gorenstein, D.G. Editor., In "^{31}P Principles and Applications" Academic Press. (1984).

[36] Gorenstein, D.G., Lai, K. and Shah, D.O. *Biochemistry* (1984) *23*, 6717.

[37] Shah, D.O., Lai, K. and Gorenstein, D.G. *Biochemistry* (1984a) *23*, 6717-6723.

[38] Shah, D.O., Lai, K. and Gorenstein, D.G. *J. Am. Chem. Soc.* (1984b) *106*, 4302.

[39] Gorenstein, D.G. Findlay, J.B., Momii, R.K., Luxon, B.A. and Kar, D. *Biochemistry* (1976) *15*, 3796.

[40] Gorenstein, D.G., Kar, D. and Momii, R.K. *Biochem. Biophys. Res. Commun.* (1976a) *73*, 105.

[41] Gorenstein, D.G., Kar, D., Luxon, B.A. and Mommi, R.K. *J. Am. Chem. Soc.* (1976b) *98*, 1668.

[42] Gorenstein, D.G., Luxon, B.A., Goldfield, E. and Vegeis, D., *Biochemistry* (1982a) *21*, 580.

[43] Gorenstein, D.G. *Bull.Magn.Reson.* (1983a) *5*, 161.

[44] Gorenstein, D.G. *Prog. in NMR Spectros.* (1983b) *16*, 1-98.

[45] Gorenstein, D.G. In "Jerusalem Symposium, NMR in Molecular Biology", (B. Pullman, Ed.,) pp. 1-15, D. Reidel. (1978).

[46] Gorenstein, D.G. and Luxon, B.A. *Biochemistry* (1978) *18*, 3796.

[47] Gorenstein, D.G. *Ann. Rev. Biophys. Bioeng.* (1981) *10*, 355.

[48] Gorenstein, D.G., Goldfield, E.M., Chen, R., Kovar, K. and Luxon, B.A. *Biochemistry* (1981) *20*, 2141.

[49] Ott, J. and Eckstein, F. *Biochemistry* (1985a) *24*, 2530-2535.

[50] Gorenstein, D.G., Schroder, S.A., Miyasaki, M., Fu, J.M., Roongta, V., Abuaf, P., Chang, A. and Yang, J.-C. "Biophosphates and their Analogues, Synthesis, Structure, Metabolism and Activity" (Bruzik, K.S. and Stec, W.J., Eds.,) Elsevier Press, pp. 487-502. (1987).

[51] Gorenstein, D.G., Schroder, S.A., Miyasaki, M., Fu, J.M., Jones, C., Roongta, V. and Abuaf, P. *Proceedings of the 10th International Conf. on Phosphorus Chemistry, Phosphorus and Sulfur*, (1987) *30*, 567-570.

[52] Fu, J.M., Schroeder, S.A., Jones, C.R., Santini, R. and Gorenstein, D.G. *J. Magn. Reson.* (1988) *77*, 577-582.

[53] Caruthers, M.H. *Acc. Chem. Res.* (1980) *13*, 155-160.

[54] Petersheim, M., Mehdi, S. and Gerlt, J.A. *J. Am. Chem. Soc.* (1984) *106*, 439.

[55] Connolly, B.A. and Eckstein, F. *Biochemistry* (1984) *23*, 5523-5527.

[56] Patel, D.J., Pardi, A. and Itakura, K. *Science* (1982) *216*, 581.

[57] Lai, K., Shah, D.O., Derose, E. and Gorenstein, D.G. *Biochem. Biophys. Res. Commun.* (1984) *121*, 1021.

[58] Sklenar, V., Miyashiro, P.N., H.Zon, G., Miles, H.T. and Bax, A. *FEBS Letts.* (1986) *208*, 94.

[59] Frey, M.H., Leupin, W., Sorensen, O.W. Denney, W.A., Ernst, R.R. and Wüthrich, K. *Biopolymers* (1985) *24*, 2371-2380.

[60] Bax, A. and Freeman, R.J. *J. Magn. Reson.* (1981) *44*, 542.

[61] Kessler, H., Griesinger, C., Zarbock, J. and Loosli, H.R. *J. Magn. Reson.* (1984) *57*, 331-336.

[62] Williamson, D. and Box, A. *J. Magn. Reson.* (1988) *76*, 174-177.

[63] Jones, C.R., Schroeder, S.A. and Gorenstein, D.G. submitted

[64] Cheng, D.M., Kan, L. and Ts'o. P.O.P. in "Phosphorus NMR in Biology", Vol. (ed. C.T. Burt) CRC Press, Inc.: Boca Raton, (1987) pp 135-147.

[65] Ott, J. and Eckstein, F. *Nucleic Acids Research* (1985b) *13*, 6317-6330.

[66] Patel, D.J. *Biochemistry* (1974) *13*, 2396-2402.

[67] Patel, D.J. *Accts. Chem. Res.* (1979) *12*, 118.

[68] Seeman, N.C., Rosenberg, J.M., Suddath, F.L., Park Kim, J.J. and Rich, A. *J. Mol. Biol.* (1976) *104*, 142-143.

[69] Lerner, D.B. and Kearns, D.R. *J. Am. Chem. Soc.* (1980) *102*, 7612-7613.

[70] Costello, A.J.R., Glonek, R. and Van Wazer, J.R. *J. Inorg. Chem. Soc.* (1976) *15*, 972-974.

[71] Giessner-Prettre, C., Pullman, C., Borer, B., Kan, L.S. and Ts'o, P.O.P. *Biopolymers* (1976) *15*, 2277.

[72] Giessner-Prettre, C., Pullman, B., Ribas-Prado, F., Cheng, D.M., Iuorno, V. and Ts'o, P.O.P. *Biopolymers* (1984) *23*, 377.

[73] Gorenstein, D.G. and Kar, D. *Biochem. Biophys. Res. Commun.* (1975) *65*, 1073.

[74] Gorenstein, D.G. *J. Am. Chem. Soc.* (1975) *97*, 898.

[75] Calladine, C.R. and Drew, H.R. *J. Mol. Biol.* (1984) *178*, 343-352.

[76] Ribas-Prado, F., Giessner-Prettre, C., Pullman, B. and Daudey, J.-P. *J. Am. Chem. Soc.* (1979) *101*, 1737.

[77] Sklenar, V., Bax, A. and Zon, H.G.*J. Am. Chem. Soc.* (1987) *109*, 2221.

[78] Hogan, M., Wemmer, D., Bray, R.P., Wade-Jardetzky, N. and Jardetzky, O. *FEBS Lett.* (1981) *124*, 202-203.

[79] Buck, F., Hahn, K.-D., Zemann, W., Ruterjans, H., Sadler, J.R., Beyreuther, K., Kaptein, R., Scheek, R. and Hull, W. *Eur. J. Biochem.* (1983) *132*, 321-327.

[80] Nick, H., Arndt, K., Boschelli, F., Jarema, M.-A.C., Lillis, M., Sadler, J.R., Caruthers, M. and Lu, P. *Proc. Natl. Acad. Sci. USA*, (1982) *79*, 218-222.

[81] Scheek, R.M., Zuiderweg, E.R.P., Klappe, K.J.M., Van Boom, J.H., Kaptein, R., Ruterjans, H. and Beyreuther, K. *Biochemistry* (1983) *22*, 228.

[82] Wade-Jardetzky, N., Bray, R.P., Conover, W.W., Jardetzky, O., Geisler, N. and Weber, K. *J. Mol. Biol.* (1979) *128*, 259-264.

[83] Pilz, I., Goral, K., Kratky, O., Bary, R.P. and Wade-Jardetzky, O. *Biochemistry* (1980) *19*, 4087-4090.

[84] Boelens, R., Scheek, R.M., VanBoom, J. and Kaptein, R. *J. Mol. Bio.* (1987) 213-234.

[85] Patel, D.J. *Accts. Chem. Res.* (1979) *12*, 118.

[86] Reinhardt, C.G. and Krugh, T.R. *Biochemistry* (1977) *16*, 2890.

[87] Buck et al., *J. Biomol. Struct. and Dyn.* (1986) *3*, 0739

[88] Ptashne, M. "A Genetic Switch". Cell Press, Cambridge MA. (1986).

[89] Lehming, N., Sartorius, J., Niemoller, M., Genenger, G. v. Wilcken-Bergmann, B. and Muller-Hill. *EMBO J.* (1987) *8*, 3143.

[90] Betz, J.L. *Gene* (1986) *42*, 283-292.

[91] Betz, J.L. *J. Mol. Biol.* (1987) *195*, 495-504

[92] Ebright, R.H. In "Protein Structure Folding and Design" (Oxender, D. Ed.) Alan R. Liss, New York, 207-219. (1986).

[93] Lomonossof, G.P., Butler, P.J.G. and Klug, A. *J. Mol. Biol.* (1981) *149*, 745-760.

[94] Drew, H.R. and Travers, A.A. *Cell* (1984) *37*, 491-502

[95] Drew, H.R. and Travers, A.A. *Nucleic Acid Research* (1985) *13*, 4445.

[96] Scheffer, I.E., Elson, E.L. and Baldwin, R.L. *J. Mol. Biol.* (1968) *36*, 291-304.

[97] Viswamitra, M.A., Kennard, O., Shakked, Z., Jones, D.G., Sheldrick, G.M., Salisbury, S. and Falvello, L. *Nature* (London), (1978) *273*, 687-690.

[98] Shindo, H., Simpson, R.T. and Cohen, J.S. *J. Biol. Chem.* (1979) *254*, 8125.

[99] Kypr, J., Sklenar, V. and Vorlic-Kova, M. *Biopolymers* (1986) *25*, 1803-1812.

7. NMR Studies of Dynamic Processes and Multiple Conformations in Protein-ligand Complexes

James Feeney

7.1 Introduction

NMR spectroscopy has now become a well-established technique for measuring the rates of dynamic processes in proteins and protein-ligand complexes. These range from processes on the picosecond time scale (such as those relating to side-chain bond rotation) to very slow processes taking days (such as exchange of some amide protons buried within protein structures). Many processes at intermediate rates such as aromatic ring-flipping, hydrogen-bond breaking and dissociation rates of protein-ligand complexes are also accessible to the method. A particular case of interest is the interconversion rate between different conformational states in a protein-ligand complex. Rates of such processes can sometimes be measured directly when the lifetimes of the different states are long (> 0.1 s). Such conformational interconversions are also involved in many other dynamic processes that require conformational readjustments of the complex to take place before they can proceed, as found for example, in side-chain bond rotations, ring-flipping and hydrogen-bond breaking. In some cases, measurements of the latter processes provide information about the dynamic processes involved in the conformational interconversions.

High resolution NMR studies have also provided the most direct evidence for the existence of multiple conformations in protein-ligand complexes in solution. We have been using this approach to study complexes of antifolate drugs bound to their target enzyme dihydrofolate reductase. We have examined more than 80 complexes formed by *L.casei* dihydrofolate reductase with a variety of substrates and substrate analogues in the absence and presence of coenzymes and have observed multiple conformations in no less than 20 complexes: in each case separate spectra have been detected for the different conformational states. It seems likely that in some of the other complexes, where multiple conformations have not been detected directly, there will probably be multiple conformations existing under conditions of rapid exchange that result in a single spectrum being observed. In this paper we will review some of the NMR measurements on dynamic processes and conformational equilibria carried out on several complexes of dihydrofolate reductase.

7.2 Background

The enzyme dihydrofolate reductase catalyses the reduction of folate and dihydrofolate to tetrahydrofolate using NADPH as coenzyme.

$$\text{Folate} + \text{NADPH} + \text{H}^+ \rightleftarrows \text{7,8-dihydrofolate} + \text{NADP}^+$$

$$\text{7,8-dihydrofolate} + \text{NADPH} + \text{H}^+ \rightleftarrows \text{5,6,7,8-tetrahydrofolate} + \text{NADP}^+$$

The final product tetrahydrofolate, is an important cofactor in a number of biosynthetic processes involved in the synthesis of purines, pyrimidines and some amino acids. Dihydrofolate reductase is thus an essential enzyme in the cell and several clinically-used drugs act by inhibiting this enzyme in invasive cells [1]: typical examples are trimethoprim (I) (antibacterial), methotrexate (II) (antineoplastic) and pyrimethamine (III) (antimalarial) (Schemes 1 and 2).

SUBSTRATES | INHIBITORS

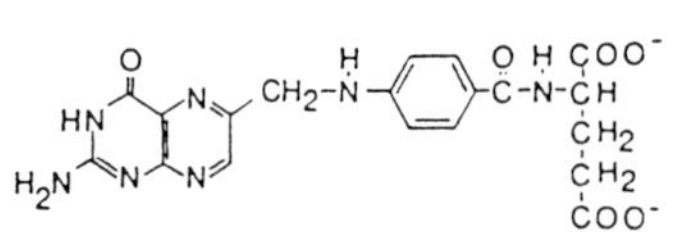

Folate

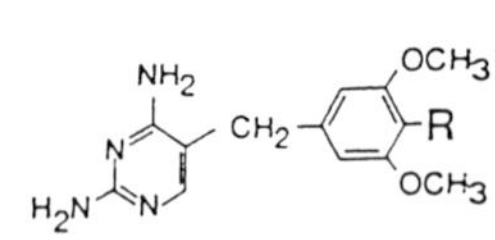

Trimethoprim R=OCH_3 I

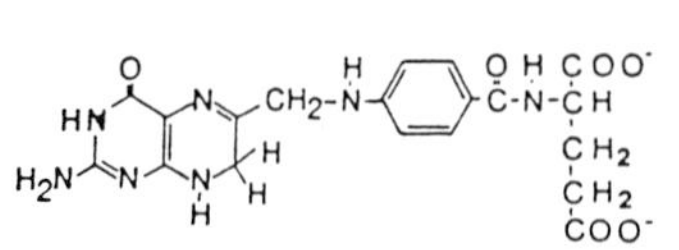

Dihydrofolate

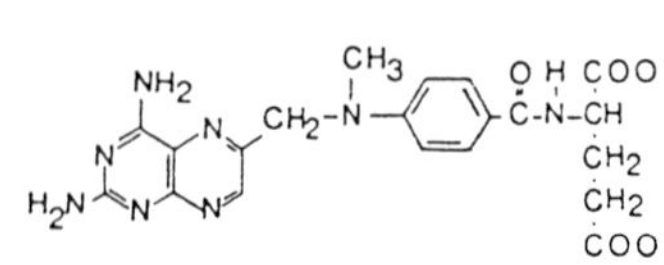

Methotrexate II

Scheme 1

In the search for improved antifolate drug molecules many analogues of these compounds have been synthesised and evaluated since the original synthesis of the parent compounds [2,3]. Detailed studies aimed at understanding the factors controlling the specificity of drug binding to the enzyme have been carried out in many laboratories, and many structural studies on complexes of dihydrofolate reductase with inhibitors have been carried out by using both X-ray [4-6] and NMR [7-13] methods. The enzymes from bacterial sources are relatively small (MW 18,000-25,000) and well-suited for NMR studies. In order to obtain detailed structural or kinetic information from the NMR studies it is first necessary to assign the NMR signals to specific nuclei in the ligand and protein. We have made such assignments for bound ligands by a combination of methods which include isotopic labelling, transfer of saturation and 2D exchange experiments [12,14-16]. Protein ^1H signals have been assigned by using selective deuteriation, 2D COSY studies and by combining data from NOESY experiments with spatial information obtained from X-ray structural studies [17,18]. Using these methods we have assigned signals for almost all nuclei in the bound coenzyme, substrates and common inhibitors and ^1H signals from about 25% of the protein residues in the *L.casei* enzyme [18]. These assigned protein signals arise from amino acids distributed throughout the structure and thus provide convenient reporter signals which can be used to monitor dynamic processes and multiple conformations in the complexes.

7.3 Dynamic Processes in Protein-Ligand Complexes

Detailed dynamic information about protein-ligand complexes has been obtained using a wide range of NMR methods including measurements of ^{13}C relaxation rates, line-shapes and intensity changes in transfer of saturation experiments. For very slow processes, rates have been measured directly by monitoring changes in signal intensities as a function of time. By using these various methods it is possible to characterize dynamic processes such as the effective overall rotation of the complex, the rapid small amplitude oscillations and rotations about specific bonds, ring-flipping of aromatic rings, breaking and reforming of hydrogen bonds and other interactions, interconversion between different conformational states, dissociation rates of the complexes and finally rates of very slow processes such as those for amide NH protons exchanging with H_2O. Clearly such dynamic information provides a more complete picture of the protein-ligand complex than that obtained from structural studies where the complex is considered as a rigid structure. This can be illustrated by examining the results of the NMR dynamic studies conducted on the complex of the antibacterical drug trimethoprim (I) and dihydrofolate reductase [19]. In this work the effective overall correlation time τ_R (15.4 (± 1.5) ns at 295 K) was estimated from ^{13}C relaxation measurements on the 7-CH_2 carbon of [7,4'-OMe$^{13}C_2$]-labelled trimethoprim bound to the enzyme: the measured value is typical of the values expected for proteins of this size (~ 18.000 MW). By analysing the relaxation rates of the 7-CH_2 and 4'-OCH_3 carbons using the Lipari and Szabo model [20] it was found that, in addition to the expected rapid rotation about the O-C bond ($> 10^{10} s^{-1}$) of the O-CH_3 group, an additional rapid motion on the subnanosecond timescale contributing to the 4'-OCH_3 ^{13}C relaxation rate must be postulated in order to explain the measured order parameter. While the data do not allow a distinction between whether this motion is a rapid oscillation about the C7-C1' or the C4'-O bond some limits on the amplitude of motion could be estimated (< ± 36°). These motions must be fast compared to $1/\tau_R$ but their precise rates cannot be estimated from this analysis.

In addition to these very fast motions there are some other much slower processes taking place within the complex that have been characterised using NMR methods. For example, the benzyl ring of bound trimethoprim undergoes a dynamic process corresponding to 180° "flips" about its C1'-C4' symmetry axis at a rate of 250 ± 27 s^{-1} at 298 K [16,19]. This rate was estimated from a lineshape analysis on the ^{13}C signals from [m-O$^{13}CH_3$]-trimethoprim bound to dihydrofolate reductase. At 273 K two separate ^{13}C signals were observed corresponding to two different environments for the m-OCH_3 groups as would be expected for a rigidly bound molecule (see Fig. 1). As the temperature was increased the signals broadened and eventually coalesced into a single resonance at 290 K. This behaviour is characteristic of exchange between two non-equivalent equally populated sites. This could be achieved by the trimethoprim benzyl ring rotating or flipping about its C1'-C4' symmetry axis thus rapidly interchanging the environment of the two m-OCH_3 groups and resulting in a single averaged resonance signal. Similar ring-flipping processes have been measured for aromatic rings in several proteins [21-24]. In most cases, a simple consideration of the crystal structure data indicates that if the structure remains static, the energy barriers to such rotations would be very high. This would certainly be the case for bound trimethoprim where the folded conformation of the molecule in the bound state would preclude a simple flipping of the benzyl ring because of the large steric interactions between atoms on the benzyl and pyrimidine rings. For such a ring-flipping to occur it is necessary to rotate θ_1 by at least 30° to remove these steric interactions and to allow flipping about θ_2 to take place. This would

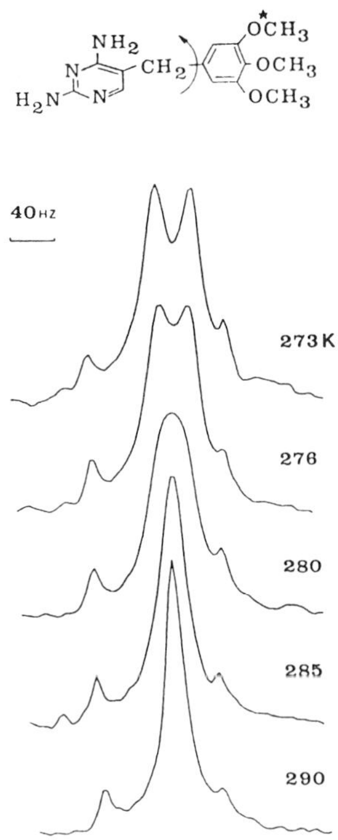

Fig. 1 The 67.9 MHz ^{13}C NMR spectra of [m-$O^{13}CH_3$] trimethoprim bound to *L.casei* dihydrofolate reductase as a function of temperature. The major resonances are those of the bound inhibitor; minor signals arise from a small amount of free trimethoprim and from natural abundance background resonance from the protein [18]. *Reprinted from Ref. 18.*

involve disrupting either the binding site of the benzyl ring or, less likely, that of the pyrimidine ring. Examination of the crystal structure data [5,25,26] indicates that the required conformational change in the ligand would need to be accompanied by a large conformational change in the protein. Thus, in measuring the rates of flipping of the benzyl ring one is indirectly monitoring the rates of the protein and ligand conformational readjustments required to allow the flipping to take place [19].

Another dynamic process which has been measured in this complex is the rate of breaking and reforming of the interaction between the pyrimidine N1 proton and the carboxylate group of Asp 26. This can be estimated from studies of the line widths of the N1H proton signal (assigned in earlier work using ^{15}N labelled trimethoprim [11]) measured over a range of temperatures. These studies, in fact, measure the exchange of the N1H proton with the protons in the H_2O solvent, a process which, in this case, is considered to be limited by the breaking of the intramolecular H-bond between the N1H proton and the Asp 26 carboxylate group. At 298 K a rate of $34 \pm 3\ s^{-1}$ was measured for this process. From the measured activation energy it seems likely that there will be conformational differences between the structures of the complexes with and without the disrupted H-bond.

Finally NMR can be used to measure the dissociation rate constant for the enzyme-trimethoprim complex. From saturation transfer measurements between ^1H signals of free and bound trimethoprim, a dissociation rate of 1.7 s^{-1} at 298 K was estimated [11]. Clearly, the previously discussed rate processes within the complex are much faster than the dissociation of the complex: during the lifetime of the complex, the benzyl ring flips at least 150 times and the N1H-hydrogen bond is broken and reformed more than 20 times. Fig. 2 summarises the kinetic processes taking place within the enzyme-trimethoprim complex. Each of these processes modulates a very important contribution to the overall binding. The lifetimes of the partially dissociated species will be very short and one might expect complete dissociation to take place only when the structure-breaking processes occur simultaneously or in very close succession. This picture supports our earlier suggestions concerning possible "zipper" type mechanisms for binding and releasing flexible ligands in protein-ligand complex formation [27]. In this model the binding of a flexible ligand to a protein is proposed to occur by the initial for-

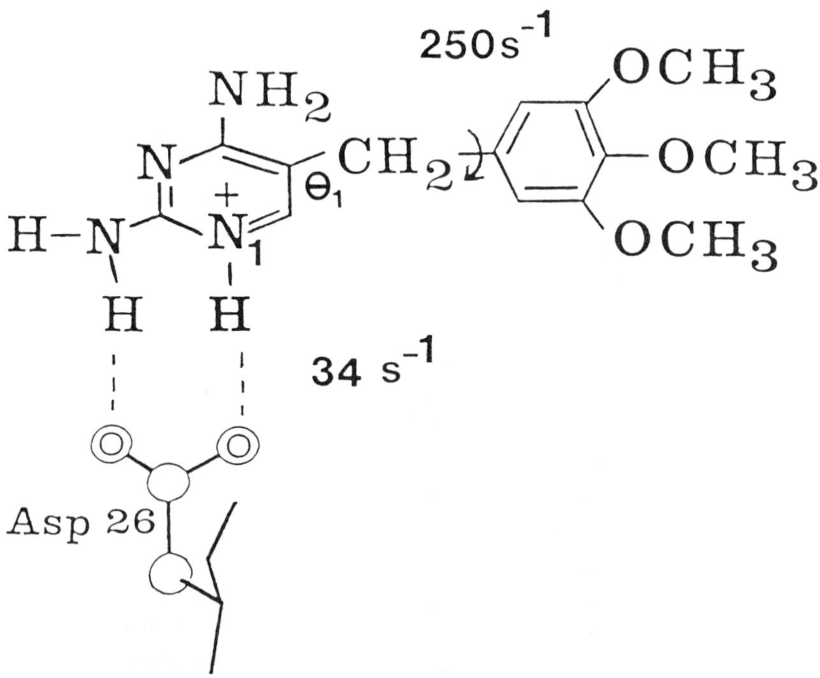

Fig. 2 Summary of the dynamic processes in the complex of *L.casei* dihydrofolate reductase and trimethoprim.

mation of a nucleation-complex where the ligand is only partially bound, which then undergoes a series of conformational rearrangements to allow the fully bound complex to form (see Fig. 3). Dissociation would proceed in a similar fashion but in reverse. Such processes would allow for rapid on-rates and off-rates while retaining the specificity of binding resulting from the multi-site interactions between the flexible ligand and protein structures.

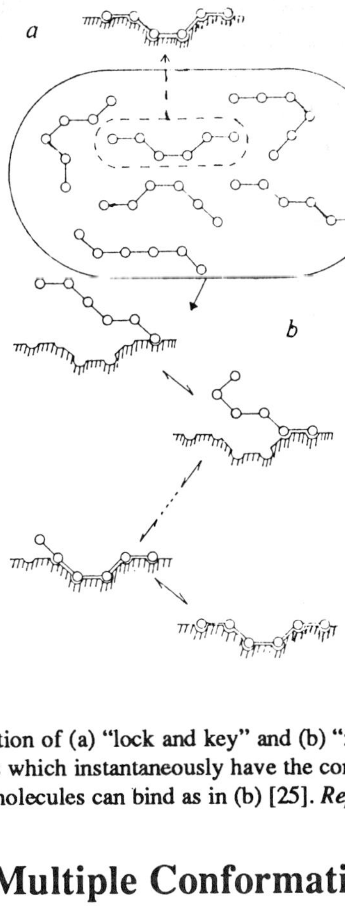

Fig. 3 Schematic representation of (a) "lock and key" and (b) "zipper" model for ligand binding. Only those ligand molecules which instantaneously have the correct conformation can bind as in (a), but essentially all ligand molecules can bind as in (b) [25]. *Reprinted from Ref. 25.*

7.4 Detection of Multiple Conformations

Early kinetic and NMR studies on dihydrofolate reductase pointed to the presence of multiple conformations in the enzyme [28-31]. The broad signals seen in the ^{1}H NMR spectrum of the *L.casei* enzyme in the absence of ligands are possibly broadened by exchange processes between different conformational states. In contrast, a complex of dihydrofolate reductase with a tightly binding ligand such as methotrexate gives a ^1H spectrum with a set of sharp signals consistent with a single conformational state. NMR studies on other complexes of the *L.casei* enzyme provide direct evidence that some of these exist as mixtures of conformations with lifetimes such that they give separate ^1H spectra. Intensity measurements provide the relative amounts of the different conformers. The conformational equilibria can often be perturbed by changing the temperature, pH or by making structural modifications to the ligand or protein (the latter by site directed mutagenesis). In some cases the rate of interconversion between the conformational states can be measured from line-shape analysis studies. Our NMR studies have uncovered three main groups of complexes showing multiple conformations, each of them being characterized by major differences in the conformations of the bound ligand in the different forms.

7.4.1 Group A. Complexes with Pyrimethamine Analogues

The NMR spectra of binary complexes of the enzyme with pyrimethamine (III) or its 4'-fluoro (IV) analogue show no indication of multiple conformations. However, the ^{19}F spectrum (shown in Fig. 4) of the complex of the enzyme with the 3'-nitro-4'-fluoropyrimethamine analogue (V) clearly shows two ^{19}F signals for (V) corresponding to two conformations A and B in the ratio of 0.6:0.4 [32]. The major difference in the conformational states appears to be in the orientation of the 3'-nitro-4'-fluoro substituted ring. By analogy with ortho substituted biphenyls one would expect that in ortho substituted phenyl pyrimidines there will

	R_1	R_2	R_3
III	C_2H_5	H	Cl
IV	C_2H_5	H	F
V	C_2H_5	NO_2	F

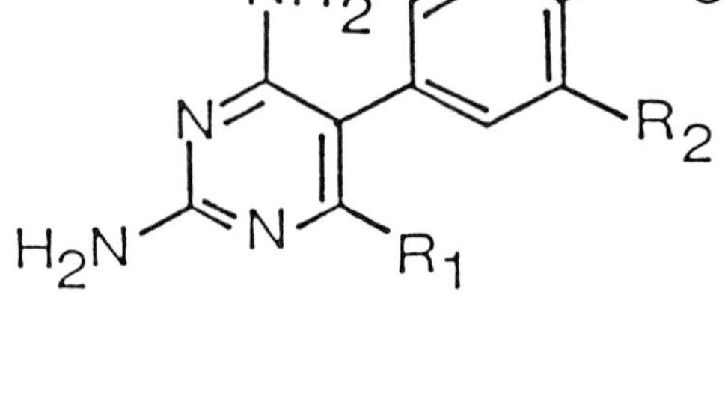

Scheme 2

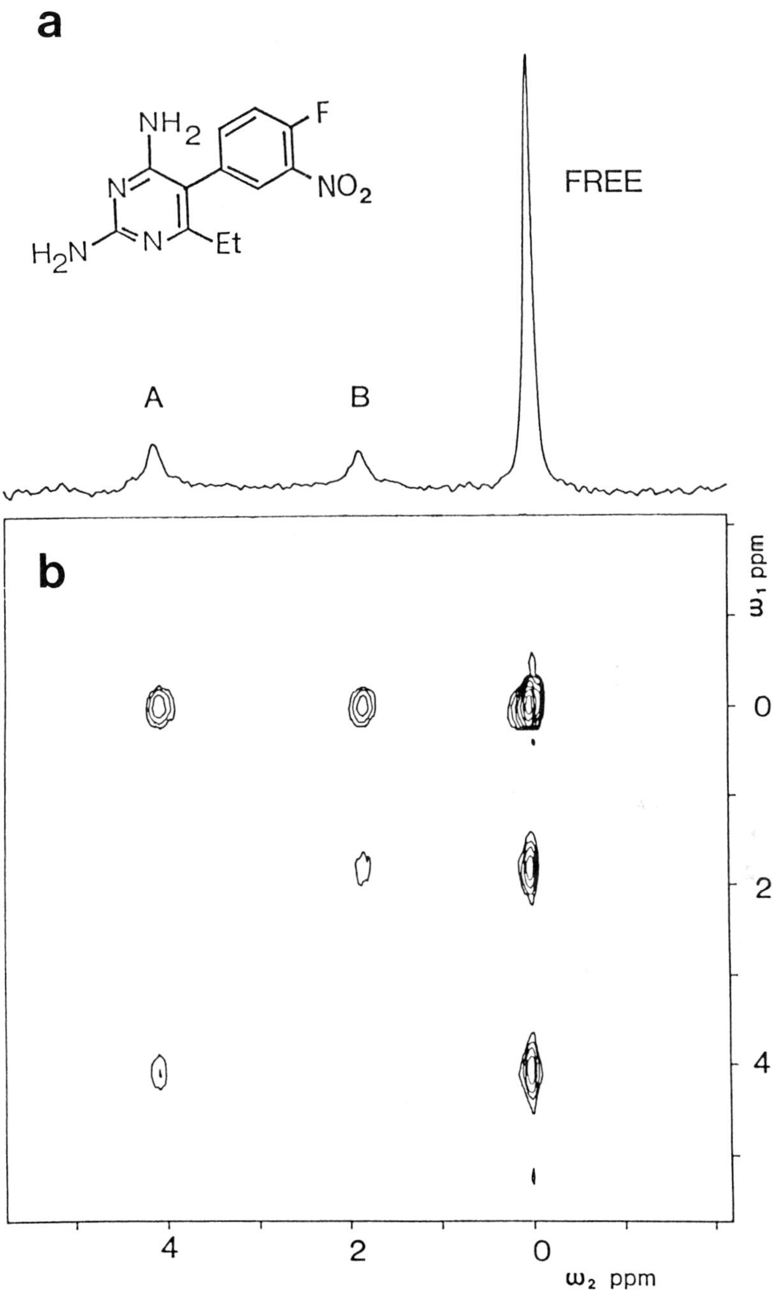

Fig. 4 376 MHz ^{19}F spectra of 1.2 mM dihydrofolate reductase in the presence of excess fluoronitropyrimethamine at 308 K, pH 6.5. (a) One dimensional spectrum, (b) Phase-sensitive 2D NOESY/Exchange spectrum (1024 data points in t_2 for each of 64 t_1 values). Both spectra are referenced to the free ligand signal [32]. *Reprinted from Ref. 32.*

be a very large energy barrier to rotation about the C5-C1' bond with the major steric interactions occurring when the two aromatic rings are coplanar. A consequence of this is that compound V exists as a mixture of slowly interconverting optical isomers in free solution and each of these binds to the enzyme. If the 2,4-diaminopyrimidine ring binds similarly in the two cases then the phenyl ring can have its 3'-nitro substituent either above or below the plane of the 2,4-diaminopyrimidine ring.

Transfer of saturation experiments and 2D exchange studies on the ^{19}F spectrum of the complex indicate that while the exchange between free and bound forms of the ligand can be detected, there is no detectable interconversion between the two bound forms themselves on the NMR timescale. This is consistent with the two bound conformations corresponding to the binding of two non-equivalent rotamers of the type indicated. Each rotamer has a different affinity for binding to the enzyme and these affinities are perturbed by the presence of $NADP^+$. Addition of coenzyme changes the specificity of binding in the different conformational states such that the ratio of forms A:B becomes 0.3:0.7 as measured from the intensity ratio of the two signals (not shown). More recent ^1H NMR studies of the binary and ternary complexes have provided further evidence for the presence of multiple conformations related to the hindered rotation in these systems [32].

7.4.2 Group B. Complexes with the Substrate Folate

NMR studies using isotopically labelled ligands, transfer of saturation and 2D exchange experiments have shown that the enzyme.folate complex exists as a mixture of at least two conformations and that the enzyme.folate.$NADP^+$ complex has three distinct conformational states (designated forms I, IIa, IIb) (see Fig. 5a and Fig. 5b) [7,12,13,33,34]. The populations of these conformational states are pH dependent. At pH 5.5 the ternary complex is almost exclusively in form I for which NOE measurements clearly indicate that the H7 proton of folate is close to the methyl group protons of Leu 19 and 27 as shown in Fig. 6. This defines the orientation of the pteridine ring in form I as being very similar to that of methotrexate in the methotrexate.enzyme complex. This orientation had previously been shown to be different from that of folate in catalytically active folate complexes [35]. One of the other forms of the enzyme.folate.$NADP^+$ complex (form IIb) does not have the same orientation of the pteridine ring as in the methotrexate complex (no NOE connections between H7 and Leu 19 and

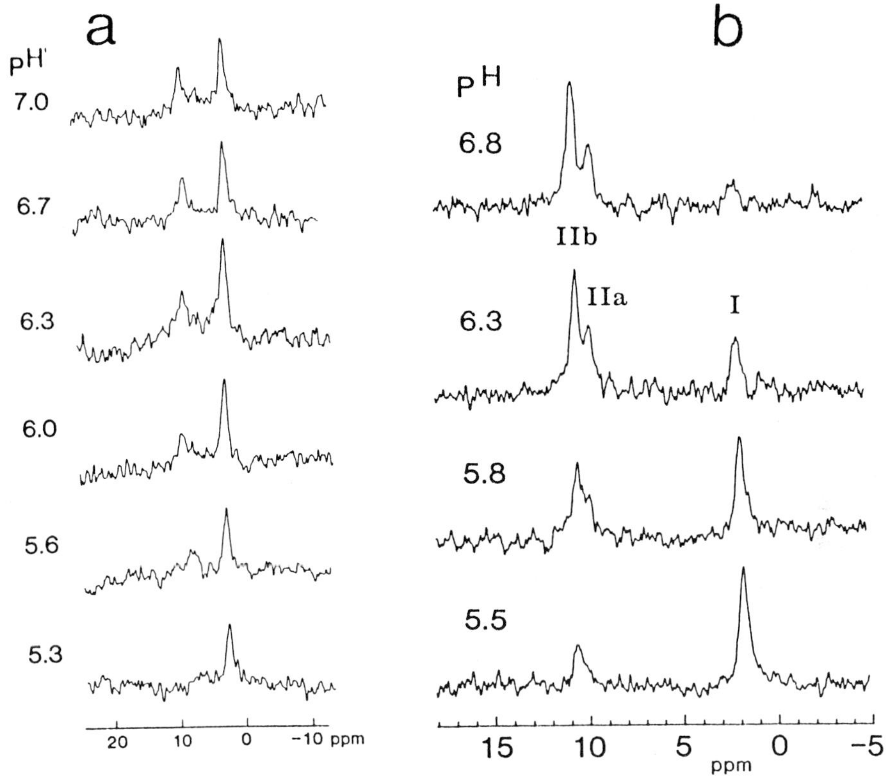

Fig. 5 (a) 20 MHz ^{15}N NMR spectra at 278 K of the dihydrofolate reductase-[5-^{15}N] folate complex at a series of pH values. (b) 20 MHz ^{15}N NMR spectra at 278 K of the dihydrofolate reductase-[5-^{15}N] folate-NADP$^+$ complex at a series of pH values [13]. *Reprinted from Ref. 13.*

27 methyl protons). The major conformational difference between forms I and IIb is the different orientation of the pteridine ring in the two forms as illustrated in Fig. 6. While the ring occupies approximately the same binding site in the different forms it differs by a 180° orientation in the two forms. Attempts to detect the interconversion rate between forms I and IIb using multiple-site transfer of saturation experiments were not successful indicating this rate to be low (< 1 s^{-1}). It seems unlikely that such an interconversion is directly implicated in the catalytic process. The similar COSY ^{1}H spectra (not shown) observed at low pH (form I) and high pH (forms IIa and IIb) indicates that the protein conformation is not very different in the three forms.

In the multiple conformations seen for complexes with pyrimethamines (Group A) and folate (Group B) the bound ligands have very different conformations but the protein conformation remains largely unchanged.

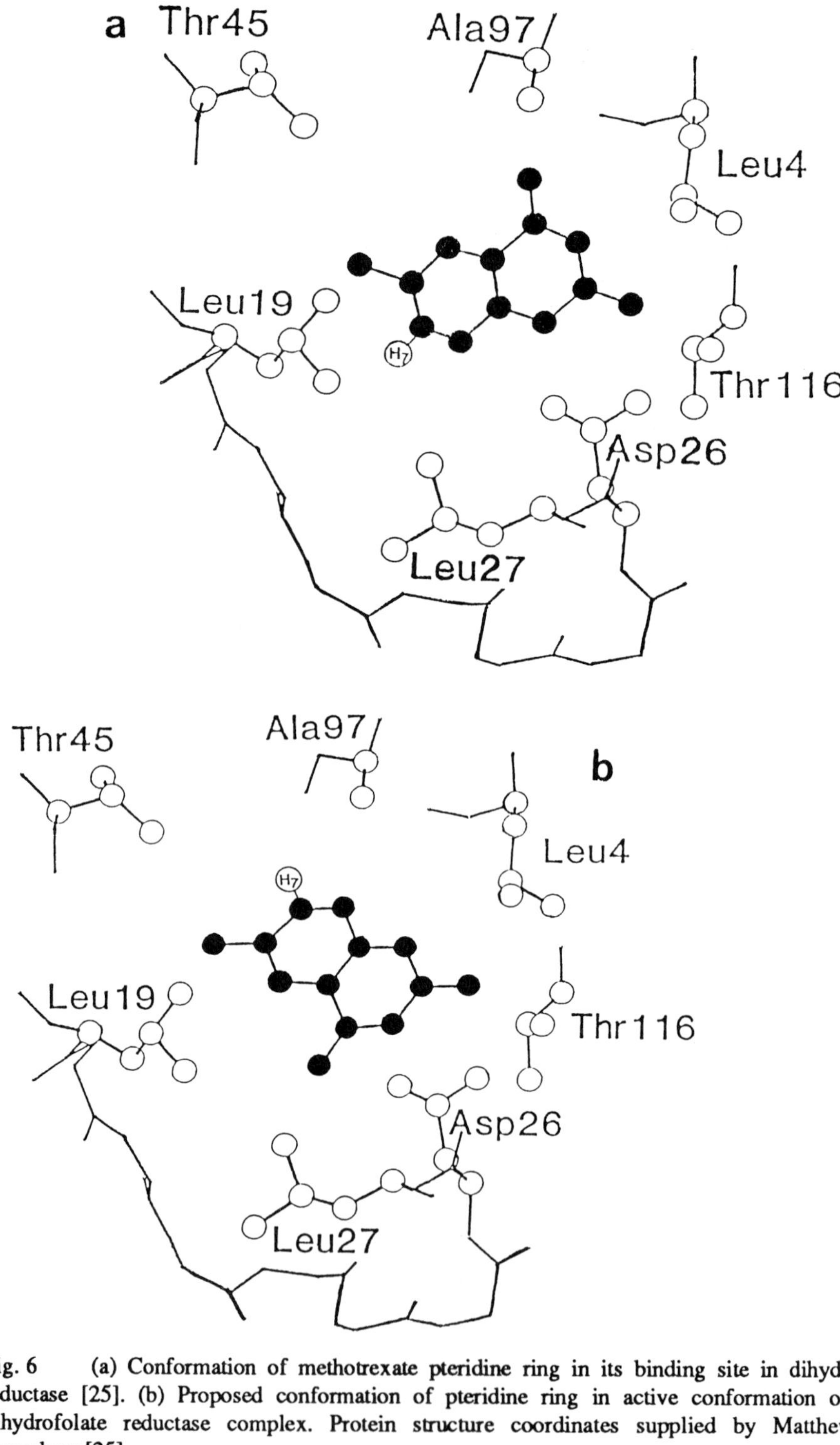

Fig. 6 (a) Conformation of methotrexate pteridine ring in its binding site in dihydrofolate reductase [25]. (b) Proposed conformation of pteridine ring in active conformation of folate dihydrofolate reductase complex. Protein structure coordinates supplied by Matthews and coworkers [25].

7.4.3 Group C. Complexes with Analogues of Trimethoprim and $NADP^+$

The ternary complex formed with trimethoprim and $NADP^+$ has been shown to exist as a mixture of two stable conformations with almost equal populations (forms I and II). The original observation of the two conformations was made in the ^1H spectrum of the complex where separate signals were detected for six of the seven histidine C2 protons in *L.casei* DHFR as shown in Fig. 7 [8,9]. On raising the temperature, the histidine "doublets" coalesced into single lines in a manner typical of a two-site exchange process: a line-shape analysis at 314 K yielded the rate of interconversion (18 s^{-1}) between the two forms. Subsequent ^{13}C and ^{31}P studies confirmed the presence of two conformations (see Fig. 8). The ^{31}P spectra are particularly useful for estimating the amounts of the two conformations. Studies of complexes of analogues of TMP and $NADP^+$ with DHFR indicated that the two conformations exist in many related complexes, the actual populations depending on the particular structures of the analogues. Such information provides an additional insight into structure/activity relationships.

A consideration of the chemical shift, coupling constant and NOE information indicates that the conformation of bound $NADP^+$ is very different in the two

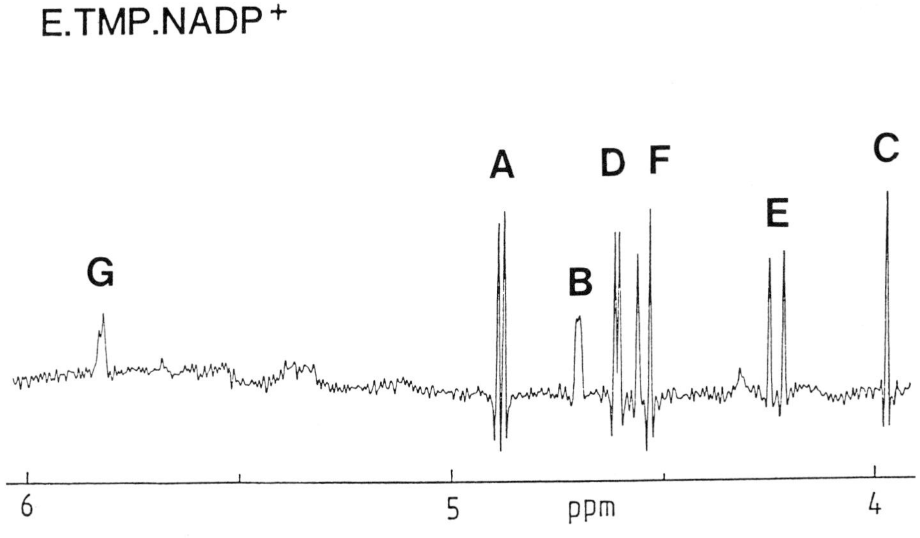

Fig. 7 Part of the deconvoluted 500 MHz ^1H NMR spectrum of the *L.casei* dihydrofolate reductase-$NADP^+$-trimethoprim complex, showing the C2-proton resonances of the seven histidine residues (labeled A-G) [12]. Chemical shifts measured from dioxane reference. *Reprinted from Ref. 12.*

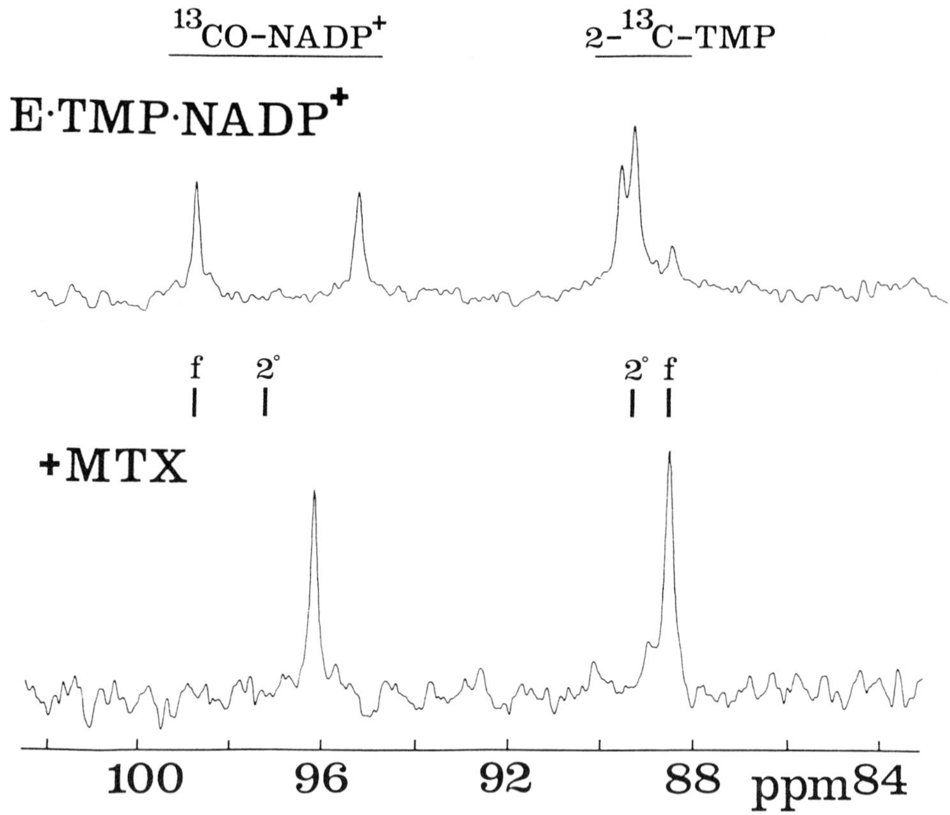

Fig. 8 (Top) The 50.3 MHz ^{13}C NMR spectra of the complex between *L.casei* dihydrofolate reductase, [carboxamide-^{13}C] NADP$^+$, and [2-^{13}C trimethoprim]. The lower field pair of resonances arise from NADP$^+$ and the higher field pair from trimethoprim (there is also a signal at 88.5 ppm from a small amount of free trimethoprim). (Bottom) The same sample after addition of excess methotrexate. The "sticks" indicate the ^{13}C chemical shifts of NADP$^+$ and trimethoprim free (f) and in their respective binary complexes (2°) [12]. *Reprinted from Ref. 12.*

forms: in form II the NADP$^+$ nicotinamide ring extends out into solution while in form I it is buried within the protein [8,9].

There are also conformational differences in the nicotinamide ribose ring and the pyrophosphate linkages between forms I and II but the adenine ring and its ribose moiety bind similarly. There is no major difference in the trimethoprim conformation between the two forms. Each of the two forms is associated with a different protein conformation and these appear to be similar in complexes formed using different analogues of trimethoprim and NADP$^+$.

The multiple conformation seen in Group A and B result mainly from the different conformations of the bound ligands. The ligands occupy essentially the same binding site in their different conformations and no substantial changes in protein conformation are detected. The multiple conformations in Group C (trimethoprim/$NADP^+$ type) not only have the ligand bound in a different conformation but part of it occupies a different binding site in the enzyme. This will result in some differences in protein conformation between forms I and II leading to differences in the ^1H protein chemical shifts of corresponding protons in the two forms.

The detection of multiconformational states in several different complexes of dihydrofolate reductase suggests that this phenomenon is probably not uncommon. One might confidently expect many examples to be uncovered as more extensive NMR studies on other protein-ligand complexes are undertaken.

Acknowledgements

I would like to acknowledge the collaboration of my coworkers in particular Berry Birdsall, Gordon Roberts, Mark Foster, Mark Searle and Saul Tendler.

References

[1] Blakley, R.L. "The Biochemistry of Folic Acid and Related Pteridine". Elsevier/North Holland, Amsterdam (1979).

[2] Seeger, D.R., Consluich, D.B., Smith, J.M. and Hultquist, M.E., *J. Am. Chem. Soc.* (1949) *71*, 1753-1758.

[3] Roth, B. and Cheng, C.C. *Prog. Med. Chem.* (1982) *19*, 1-58.

[4] Matthews, D.A., Alden, R.A., Bolin, J.T., Filman, D.J., Freer, S.T., Hamlin, R., Hol, W.G.J., Kisliuk, R.L., Pastore, E.J., Plante, L.T., Xuong, N. and Kraut, J., *J. Biol. Chem.* (1978) *253*, 6946.

[5] Bolin, J.T., Filman, D.J., Matthews, D.A., Hamlin, R.C. and Kraut, J. *J. Biol. Chem.* (1982) *257*, 13650.

[6] Kuyper, L.F., Roth, B., Baccanari, D.P., Ferrone, R., Beddell, C.R., Champness, J.N., Stammers, D.K., Dann, J.G., Norrington, F.E.A., Baker, D.J. and Goodford, P.J., *J. Med. Chem.* (1982) *25*, 1120-1122.

[7] Birdsall, B., Gronenborn, A., Clore, G.M., Roberts, G.C.K., Feeney, J. and Burgen, A.S.V., *Biochem. Biophys. Res. Commun.* (1981) *101*, 1139.

[8] Gronenborn, A., Birdsall, B., Hyde,.E.I., Roberts, G.C.K., Feeney, J. and Burgen, A.S.V., *Nature* (1981) *290*, 273.

[9] Gronenborn, A., Birdsall, B., Hyde, E.I., Roberts, G.C.K., Feeney, J. and Burgen, A.S.V., *Molecular Pharm.* (1981) *20*, 145.

[10] Cayley, J., Albrand, J.P., Feeney, J., Roberts, G.C.K., Piper, E.A. and Burgen, A.S.V., *Biochemistry* (1979) *18*, 3886.

[11] Bevan, A.W., Birdsall, B., Gronenborn, A., Potterton, E., Clore, G.M., Roberts, G.C.K., Feeney, J. and Burgen, A.S.V. "Pteridines and Folic Acid Derivarives" Blair, J.A., Ed. de Gruyter, Berlin (1983).

[12] Birdsall, B., Bevan, A.W., Pascual, C., Roberts, G.C.K., Feeney, J., Gronenborn, A., and Clore, G.M., *Biochemistry* (1984) *23*, 4733.

[13] Birdsall, B., De Graw, J., Feeney, J., Hammond, S., Searle, M.S., Roberts, G.C.K., Colwell, W.T. and Crase, J., *FEBS Letters* (1987) *217*, 106.

[14] Birdsall, B., Gronenborn, A., Hyde, E.I., Clore, G.M., Roberts, G.C.K., Feeney, J. and Burgen, A.S.V., *Biochemistry* (1982) *21*, 5831.

[15] Hyde, E.I., Birdsall, B., Roberts, G.C.K., Feeney, J. and Burgen, A.S.V., *Biochemistry* (1980) *19*, 3738.

[16] Cheung, H.T.A., Searle, M.S., Feeney, J., Birdsall, B., Roberts, G.C.K., Kompis, I. and Hammond, S.J., *Biochemistry* (1986) *25*, 1925.

[17] Hammond, S.J., Birdsall, B., Feeney, J., Searle, M.S., Roberts, G.C.K. and Cheung, H.T.A., *Biochemistry* (1987) *26*, 8585.

[18] Hammond, S.J., Birdsall. B., Searle, M.S., Roberts, G.C.K. and Feeney, J., *J. Mol. Biol.* (1986) *188*, 81.

[19] Searle, M.S., Forster, M.J., Birdsall, B., Roberts, G.C.K., Feeney, J., Cheung, H.T.A., Kompis, I. and Geddes, A.J. *Proc. Natl. Acad. Sci., U.S.A.* (1988) *85*, 3787.

[20] Lipari, G. and Szabo, A., *J. Am. Chem. Soc.* (1982) *104*, 4546-4570.

[21] Campbell, I.D., Dobson, C.M. and Williams, R.J.P., *Proc. Roy. Soc. Lond. B.* (1975) *189*, 503.

[22] Campbell, I.D., Dobson, C.M., Moore, G.R., Perkins, S.J. and Williams, R.J.P. *FEBS Letters* (1976) *70*, 96.

[23] Snyder, G.H., Rowan III, R., Karplus, S. and Sykes, B.D. *Biochemistry* (1975) *14*, 3765.

[24] Wüthrich, K. and Wagner, G. *FEBS Letters* (1975) *50*, 265.

[25] Matthews, D.A., Bolin, J.T., Burridge, J.M., Filman, D.J., Volz, K.W., Kaufman, B.T., Beddell, C.R., Champness, J.N., Stammers, D.K. and Kraut, J. *J. Biol. Chem.* (1985) *260*, 381.

[26] Baker, D.J., Beddell, C.R., Champness, J.R., Goodford, P.J., Norrington, F.E.A., Smith, D.R. and Stammers, D.K. *FEBS Lett.* (1981) *126*, 49.

[27] Burgen, A.S.V., Roberts, G.C.K. and Feeney, J., *Nature* (1975) *253*, 753.

[28] Dunn, S.M.J., Batchelor, J.G. and King, R.W., *Biochemistry* (1978) *17*, 2356.

[29] Pattishall, K.H., Burchall, J.J. and Harvey, R.J. *J. Biol. Chem.* (1976) *251*, 7011.

[30] London, R.E., Groff, J.P. and Blakley, R.L., *Biochem. Biophys. Res. Commun.* (1979) *86*, 779.

[31] Cayley, P.J., Dunn, S.M.J. and King, R.W., *Biochemistry* (1981) *20*, 874.

[32] Tendler, S.J.B., Griffin, R.J., Stevens, M.F.G., Birdsall, B., Roberts, G.C.K. and Feeney, J. *FEBS Letters* (1988) *240*, 201.

[33] Birdsall, B., Feeney, J., Tendler, S.J.B., Hammond, S.J. and Roberts, G.C.K., *Biochemistry* (1989) *28*, 2297.

[34] Birdsall, B., Andrews, J., Ostler, G., Tendler, S.J.B., Feeney, J., Roberts, G.C.K., Davies, R.W. and Cheung, H.T.A., *Biochemistry* (1989) *28*, 1353.

[35] Charlton, P.A., Young, D.W., Birdsall, B., Feeney, J. and Roberts, G.C.K., *J. Chem. Soc. Chem. Commun.* (1979) 922.

Index